To all the teachers who inspired us, encouraged us,
and supported us throughout our lives.

THE SCIENCE OF

ALFRED HITCHCOCK

THE TRUTH BEHIND PSYCHO, THE BIRDS, NORTH BY NORTHWEST, AND OTHER LEGENDARY FILMS BY THE MASTER OF SUSPENSE

BY MEG HAFDAHL & KELLY FLORENCE

Skyhorse Publishing

Skyhorse Publishing books may be purchased in bulk at special discounts for sales promotion, corporate gifts, fund-raising, or educational purposes. Special editions can also be created to specifications. For details, contact the Special Sales Department, Skyhorse Publishing, 307 West 36th Street, 11th Floor, New York, NY 10018 or info@skyhorsepublishing.com.

Skyhorse® and Skyhorse Publishing® are registered trademarks of Skyhorse Publishing, Inc.®, a Delaware corporation.

Visit our website at www.skyhorsepublishing.com.

10 9 8 7 6 5 4 3 2 1

Library of Congress Cataloging-in-Publication Data is available on file.

Cover design by David Ter-Avanesyan

ISBN: 978-1-5107-7964-8
Ebook ISBN: 978-1-5107-7965-5

Printed in the United States of America

Table of Contents

Introduction

In an episode of *Alfred Hitchcock Presents*, the legendary director, host of the show, quipped, "I think everyone enjoys a nice murder, provided he's not the victim." This philosophy got Alfred Hitchcock far. It was his darkly comic take on suspense, his cheeky public persona, and, most important, his tenacious auteurism that made for an unforgettable career. While Hitchcock made enemies along his path (you'll learn about some of them in this book), he also endeared the world to him, with his wit, style, and unique ability to make us both laugh and scream. When the studio and film-rating heads were horrified by too much skin in the shower scene in *Psycho*, Hitchcock pretended to reedit it, fooling them into thinking he'd taken out some of Janet Leigh's nude body. He hadn't. They believed him, and the movie is cinematic history. When the National Park Service was furious over a fictional murder and hi-jinks happening at a reconstructed Mount Rushmore in *North by Northwest*, the director went ahead with, again, one of the most legendary moments in film. Sometimes we need a hero who's willing to ruffle a few feathers. And Alfred Hitchcock was certainly that man. But he was also loyal to many actors and colleagues, and he made sure his wife, Alma, was by his side, approving many of his decisions.

Like all the books in our *The Science of...* series, we delve into the inner workings of not only the artist but also the art itself. We examine the psychology in thrillers such as *Rear Window* and *Marnie*, as well as biology, chemistry, and so much more. Plus, we are true fans of Alfred Hitchcock and his films, having been watching his efforts since we were little girls. It's a true honor to research his life, his influences, and the inspirations that were the catalyst for some of our favorite thrillers.

With a keen eye and an unwavering imagination, Alfred Hitchcock led the moviegoing public into the twentieth century. Now, follow us as we learn about the man, the myth, and the legend together, as well as the fascinating science behind his films.

We are aware of the accusations made against Alfred Hitchcock. We do not condone or make excuses for this behavior. We have included mention of it in this book.

CHAPTER ONE

The Lodger

A baby is born on August 13th, 1899, in the rented flat above his father's grocery on the outskirts of London. He is the youngest child of Emma and William, little brother to Ellen "Nellie" and William Jr. We'll cut out the suspense and tell you this baby grows up to be known as one of cinema's most revered filmmakers, the "Master of Suspense" himself.

Alfred Hitchcock's legacy is rooted in the horror and mystery genres, in which he proved adept at not only scaring his audiences but also delighting and thrilling them. As we've come to understand as horror aficionados, many people question what leads a person to pursue art that explores darker themes in life. Hitchcock found the answer in his childhood, in which his loving but strict parents would dole out rather extreme punishments. Hitchcock often recounted one incident to friends and interviewers: "William Hitchcock reportedly taught his son a lesson at a tender age, sending Alfred off to the local police station with a note that said the boy had been naughty. The policeman locked him in a cell, telling him, 'This is what we do to naughty boys.' Hitchcock said he always remembered the clang of the door, 'which was the potent thing—the sound, and the solidity of that closing cell door and the bolt.'"[1] After about five or ten minutes, the officer let little Alfred go, forever traumatizing him. Hitchcock claimed to be afraid of police and prisons into his adulthood.

> Having a fear of the police is also known as capio-phobia, from the Latin word *capio*, meaning "arrest," and the Greek *phobos*, for "fear." Like most phobias, this fear can be treated with mental-health therapy, medications, or a combination of the two.[2]

Although this story could explain, in part, why Hitchcock became a purveyor of plots centered on crime and suspicion, many question if it even happened. His sister, Nellie, remembered her brother's brief incarceration, but over the years Alfred Hitchcock's telling of the incident changed. He would tell interviewers and friends different versions; in some he was aged four, whereas he told others he was closer to eleven. Perhaps that is the sign of a genius filmmaker—someone who knows how to enhance a story for maximum effect.

Raised in a devout Catholic household, Hitchcock was educated in religious schools and had many memories of Sunday Mass with his family. Thanks to William Hitchcock's successful grocery (and two fish-and-chips shops), Alfred and his siblings grew up with creature comforts, like books. A particular favorite was *Grimm's Fairy Tales*. Alfred Hitchcock became entranced with its suspenseful storytelling. Perhaps it was the books of his youth that compelled Alfred, known as "Alfie" to his friends, to choose an adventurous course of career. At twelve years old, having graduated from his formal religious education, Hitchcock chose to go to school to be a navigator. Through courses such as magnetism and chemistry, he became well versed in engineering.

As you might have guessed, navigation and engineering didn't end up holding great appeal for Hitchcock. Following two profound moments in his young life—the death of his father from a heart condition when Alfie was only fifteen and the devastation of London in World War I—Alfred chose another avenue: art.

Hitchcock changed his career trajectory by signing up for art classes at Goldsmiths College, a progressive branch of London University. He was tasked to sketch people in the real world, giving him his first taste of using visual art to tell human stories.

According to Hitchcock scholar Patrick McGilligan, "It was at Goldsmiths that Hitchcock first began to pay attention to the history and principles of art: composition, depth of field, the uses of color, shadow and light. He began to frequent art galleries and museums, especially entranced by the French moderns. Art courses sharpened his interest in theater and film. Hitchcock now became an inveterate 'first-nighter' and the West End plays he saw during the years before he entered the film industry and in the 1920s made a lasting impression."[3] One of those plays was *The Lodger*, which would later be adapted to the silver screen and become Alfred Hitchcock's first released film.

In 1921, because of his background in art at Goldsmiths College, Hitchcock was offered a position at British Famous Players–Lasky as head of the art title department. As he recalled, "They offered me seven pounds a week, but I insisted that was too much, asked for less and told them to give me a raise if it worked out later."[4] We're going to guess Hitchcock went on to make a much higher salary!

Rising in the ranks at the studio, Alfred Hitchcock directed three silent films before *The Lodger*, two of which had been shelved. Before *The Lodger*, his film *The Pleasure Garden* was released in 1925. Though it marks his first feature-length film, it doesn't hold the same trademark dark content and suspense as *The Lodger*.

Although Hitchcock had a few films under his belt, the studio perceived him as inexperienced, so Famous Players–Lasky (in partnership with Gainsborough Pictures) hired a more seasoned editor, Ivor Montagu, to reedit *The Lodger*. In a 1980 article for *Sight and Sound* magazine, Montagu recounted his initial reaction to *The Lodger*: "They ran the film, with which at once I fell enthusiastically in love. Now, the hackneyed treatment of the plot and a weakness in characterization make it look primitive. Then, by contrast with the work of his seniors and contemporaries, all Hitch's special qualities stood out raw: the narrative skill, the ability to tell the story and create the tension in graphic combination, and the feeling for London scenes and characters."[5] "Hitch" was an obvious talent from the start.

The Lodger: A Story of the London Fog is based on the 1913 novel *The Lodger*, by British writer and vocal suffragist Marie Belloc Lowndes. Often compared to her more well-known contemporary Agatha Christie, Lowndes wrote on similar themes of murder, mystery, and intrigue. The suspenseful plot of the novel translated well on film; it was first adapted by Hitchcock in 1927, as well as three more times by other directors, most recently in 2009. We guess murder never goes out of style.

The Lodger is considered the first book to be inspired by the unsolved Whitechapel case of 1888, or more notably known as the Jack the Ripper murders. (If you want to know more about Jack's reign of terror, pick up a copy of our 2021 book *The Science of Serial Killers*.) We were able to watch *The Lodger* on YouTube, which we'll assume is not how Hitchcock would've preferred we screen his first film. If we were alive in 1927, we would've dressed in our best to attend *The Lodger*. It would've been shown in a theater that was constructed to look like its European counterparts, full of lush seating, velvet drapes, and gold accents. While the first silent films in the 1910s were indeed silent, over the next decade live music accompanied films. In 1927 any number of instruments, from organs to full orchestras, could've been set up to play along with the movie. For *American Symphony,* John Pruitt described the fad:

> From a strictly musical point of view, the trend towards a complex and sensitive approach to film accompaniment was, not surprisingly, a gradual, evolutionary process. The first projected films (i.e., not the "peep shows") were accompanied by a solo piano player who usually improvised, often mixing snatches of popular songs and passages from "the classics." Eventually, films were distributed with published cue sheets suggesting what the piano player (or possibly an organist) would ideally perform in this same "checkerboard" fashion. By the 1920s, a major film released in a large city might be accompanied by a full orchestra, but since orchestras can't improvise, complete scores became a necessity. Yet again, these were most often a mixture of current hit tunes and classical favorites. Conductors who specialized in cinematic accompaniment would compile the scores and usually compose original "bridge" material themselves since synchronization was an immense tricky problem.[6]

This partnership of music and cinema brought theater, opera, and ballet patrons a new way to enjoy entertainment—the same magical storytelling still sought one hundred years later, thanks to famous movie composers like John Williams, Danny Elfman, and Holly Amber Church, whom we interview in chapter fifteen. As we touch on through this book, the iconic music of Alfred Hitchcock's films endures, even if we couldn't fully appreciate the grandeur of live music as we watched *The Lodger* on our iPads.

One special note that many Hitchcock fans have come to love about the iconic director: he makes his first cameo in *The Lodger*. Said to be inspired by filmmaker D. W. Griffith, Hitchcock is known to make small appearances in all his films. The silent *The Lodger* is no exception. Blink or you'll miss his back to the camera. It's not as memorable as his cameos in *Psycho* or *The Birds*, but it's the first!

Now, to the science. In the opening sequence of the film, the police are seen giving a distraught eyewitness a sip of a beverage to calm her nerves. Why does this work? First, offering a sip of water or other liquid can serve as a distraction in the moment. It takes effort to shift focus and drink and this slows down a person's breathing. Second, water has been proven to reduce anxiety, lower depression, and increase happiness. A 2015 study found that dehydration affects mood, fatigue, and levels of alertness.[7] Try taking a sip of water the next time you're feeling anxious, and see if it helps!

The closeup shots of people reacting to the news of the murder in *The Lodger* clearly show the fear and shock on their faces. Why are some emotions so easy to read? Wherever we are born on the planet, we all share the same primary emotions: fear, happiness, sadness, disgust, surprise, and anger. As infants, we naturally display these feelings, and, if we are emotionally and socially intelligent, we are able to read these emotions in others. It's not until we get older that we are taught to emphasize or hide certain facial expressions. For example, you may be taught from a young age to pretend to like the food you're given when visiting someone's house or feign happiness when receiving a gift, perhaps hiding your true feelings. In some cultures, kids are encouraged not to show anger or sadness openly, but in others they are scolded for showing too much happiness. Clearly, especially in the silent-movie era, showing emotions openly is important for the actors to get the message across to the audience.

The serial killer in *The Lodger* has a type: blonde women. Have serial killers targeted people with specific traits in the past? Ted Bundy, whom we wrote about in *The Science of Serial Killers* (2021), was known to go after young, brunette women, and the "Son of Sam," David Berkowitz, was fixated on attractive women with long, dark hair. Experts believe serial killers share common traits, including "sensation seeking, a lack of remorse or guilt, impulsivity, the need for control, and predatory behavior."[8] Although many killers in media have been portrayed as charming and manipulative, it's important to note that not all share the same characteristics.

In *The Lodger*, as the news gets out about the killer targeting blondes, one of the models declares, "No more peroxide for yours truly!" How does peroxide work? According to the website StyleCraze, hydrogen peroxide tones down or lightens natural hair color by oxidizing the hair pigment (melanin) and keratin (the hair's structural protein).[9] Although bleaching or dying hair seems fairly common these days, it surprised us to learn that evidence exists that people have been coloring their hair since 1500 BCE! Egyptians, Greeks, and Romans used items like henna and plant extracts to darken their locks. Around 300 BCE hair color was used to signify rank and profession, and it was used to intimidate on the battlefield.[10] It wasn't until the 1800s that a chemist, William Henry Perkin, made an accidental discovery that remains a staple of hair dyes to this day. While trying to cure malaria, Perkin created the first synthesized dye in which a color-changing molecule was derived. By synthesizing quinine, it created a mauve-hued substance. The color, named mauveine, went on to be patented by Perkin and was popular in fashion for several years during this time.[11]

> In 1931 Howard Hughes released a film called *Platinum Blonde*, titled to promote and capitalize on the hair color of the young star Jean Harlow. Many fans quickly followed suit, dyeing their hair to match Harlow's.[12]

When the mysterious lodger shows up to the property in the film, he appears to be ill at ease. Noises frighten him, and he seems annoyed at things happening around him. While we learn later that he has a past and a plan of his own that he's concentrating on, some might interpret his actions as cold. Even though these are writing, directing, and acting choices, real conditions affect people's interpretation of their surroundings. Sensory processing disorder (SPD) affects how the brain processes sensory information. Too much noise, bright lights, and excessive visual stimuli can all be disruptive to those with SPD. Strong scents, tastes, and even textures can affect people as well. According to the National Institute of Health, an estimated 5 percent to 16.5 percent of the general population has symptoms associated with sensory processing disorder.[13] If you or someone you know is showing signs of SPD, it's important to work with an occupational therapist to gain skills to help cope. Having a quiet place or a time set aside to avoid stimuli is also recommended.

Before the discovery of the Avenger's eighth murder in *The Lodger*, Daisy's parents are seen yawning back and forth. Speculations about yawning have been debated since the days of Hippocrates, who believed yawning was related to having a fever. In modern times, scientists have concluded that it is a normal human behavior that we can't control. A professor of obstetrics discovered that the human fetus yawns during its first trimester in the womb![14] We tend to yawn when we feel sleepy or bored, but it's more of a signal to our brains to stay alert. Why does yawning seem to be contagious? One theory is the phenomenon known as echopraxia in which we see a certain behavior and mimic it. This also explains how certain emotions can be contagious. We may mirror someone's silly, giddy mood and get the giggles ourselves, while other days, someone's snarky attitude puts us in a bad mood. Hopefully the mystery, romance, and surprises put you in a good mood after reading about *The Lodger* or at least intrigued you enough to read on.

CHAPTER TWO

Blackmail

Before Alfred Hitchcock transitioned to screenwriting and directing, he made a splash in the art department of Famous Players–Lasky. He'd had experience creating visuals for the W. T. Henley's Telegraph Works Company's advertising, where he'd been employed for five years. It was his work in ads and his art portfolio that impressed the studio heads. This led to his designing of intertitle cards for a dozen silent films. "Silent film intertitles were the written texts that were placed on the screen to help the audience connect and understand the action taking place. . . . They were text that would intercut the action on the screen to deliver pertinent information about the film's plot or action to the audience."[1] Some intertitles were plain—think black-and-white text—while others had decorative flourishes that artists like Hitchcock would design. Unfortunately, these twelve films, including titles like *Three Live Ghosts* (1922) and *The Spanish Jade* (1922), are tough to find in streaming, as many silent films weren't preserved properly. In America, there are known to be over seven thousand lost silent films from 1912 to 1929 due to improper storage, neglect, or other factors.[2]

> To become a film archivist, you'll need to obtain a bachelor's degree, gain experience through internships, and get a master's degree in archival studies or library science for advancement.[3]

After his stint creating intertitles, Hitchcock then went on to be the art director of six films in the early 1920s, several years before he directed *The Lodger*. He often filled in other roles, such as assistant directing or adapting novels to the screen. But many believe it was his time spent in art direction that made for his incredible ability to create such stunning compositions on film. In his piece for *Print*, Hitchcock biographer Donald Spoto wrote, "Hitchcock's keen eye for detail, his understanding of the components of a powerful visual image and his strong sense of composition and design were instantly recognized and sharpened his desire to direct his own feature."[4] Because of his experience in so many aspects of filmmaking, Hitchcock was able to have step-by-step control from script to storyboard to camera. Whereas other directors often allowed designers to create, Hitchcock was known to have every aspect of the scene already planned. His process was outlined by Spoto: "When shooting script and storyboard are complete and casting concluded, Hitchcock begins conferences with costume and set designer, art director and set dresser, composer, and credit designer. 'It's all in the script, dear Edith,' (Hitchcock) remarked to Edith Head, who has designed costumes for many of his films. The shooting script for *Vertigo* (1958)—arguably his masterpiece—specifically states that Kim Novak is to be dressed in a pale gray suit so that, with her blonde hair, 'she looks as if she just stepped out of the San Francisco fog.'"[5]

Now, let's rewind the clock back to before Hitchcock was an icon, when he was still getting his feet proverbially wet in filmmaking. *Blackmail* is his first picture with sound, although he shot it as a silent film. In 1929, movies were transitioning into talkies. Because sound was a new technology, not all theaters were equipped, so there are two versions of *Blackmail*: one with the classic intertitles and one with dialogue. If you were a moviegoer in the late twenties, where you lived and how sophisticated the theater was would dictate which version of *Blackmail* you saw. The one with dialogue is the one we were able to watch on the streamer Plex, and we were pleased with how well preserved the film is. It is easy to pick up on the silent-film conventions of acting style, exposition, and more in *Blackmail* because it was originally intended to not have sound. When it became clear that talkies were the new wave, the studio had Hitchcock reshoot scenes with dialogue. This proved challenging because the film's star, Anny Ondra, was a Czech actress with a strong accent that didn't match her British character, Alice White, girlfriend to a detective at the Scotland Yard. Therefore, Ondra was tasked with mouthing the words, while British actress Joan Barry recorded the sound. This odd amalgam of silent and talkie is described by Fritzi Kramer on the blog *Movies Silently*:

> Part-talkies have not aged well because they come off as the stop-gap cash grab that they were. Adding a reel or two of sound to an otherwise silent film allowed producers to bill the film as a "talking picture" without completely reshooting it. However, the technique also left lurching celluloid Frankenstein monsters that are not particularly enjoyable to fans of either silents or early talkies. Hitchcock produced something a little more sophisticated instead. It's not a complete talkie but it's not a part-talkie either. Some scenes play silent with synchronized sound and no intertitles, which creates a surreal atmosphere. Other scenes were completely reworked with sound in mind and rather cleverly too.[6]

Even though it didn't start out that way, *Blackmail* became the first British film to have the honor as a full talking picture.

> Contemporary silent films include *Tuvalu* (1999) and *The Artist* (2011), which won Best Picture at the Academy Awards.

As you might've guessed from the title, *Blackmail* holds all the intrigue, crime, and suspense that would later become Alfred Hitchcock's signature. It's the story of a jilted girlfriend, Alice, who makes the fateful mistake of joining Mr. Crewe, played by Cyril Ritchard, up to his artist's studio. Unfortunately, it becomes clear to Alice, and to the audience, that Mr. Crewe has more than art on his mind. The dread is palpable as he continues to make unwanted sexual overtures toward Alice, when, finally, as she tries to escape, he attempts to rape her. Honestly, we were a bit taken aback by the frankness of the scene, considering the film is nearly one hundred years old. As mentioned in *Alfred Hitchcock: A Life in Darkness and Light*, this is only the beginning of a topic that will work its way into much of his work: "Hitchcock envisioned *Blackmail* as a splendid pretext to exploring the primal commingling of sex and violence that he had already marked as his territory in *The Lodger*."[7]

Alice defends herself from Mr. Crewe with a nearby knife (go Alice!), which leaves the scoundrel dead. We watch as Alice goes into shock at what has occurred, both the trauma of the assault and the death sending her into a numb sort of hysteria. Hitchcock spends time on this aftermath, drawing out the suspense. The next morning, at breakfast, Alice is asked to cut the bread with a strikingly similar knife, which evokes another rise of panic that Hitchcock masterfully depicts using sound (the repetition of the word *knife*) and a close-up on Alice as she is awash in panic.

As the title denotes, there's a witness (Donald Calthrop) to the crime who attempts to blackmail Alice and her detective boyfriend. Without giving the plot away, we'll say he made a grave error by doing this.

> We were delighted to see that Alfred Hitchcock makes one of his famous cameos in *Blackmail*, especially since he looks so young! At only thirty, Hitchcock is certainly recognizable, yet with a bit more hair on top! He sits on a train, while a little boy (Jacque Carter) bothers him by messing with his hat. It's a nice spot of comical whimsy within a rather serious film.

Like many films of the era, *Blackmail* was based on a play. In 1928, *Blackmail* ran for nearly forty performances in the Globe Theatre in the West End of London, directed by actor Raymond Massey, who later starred in one of our favorite play-to-film adaptations, *Arsenic and Old Lace* (1944). The playwright, Charles Bennett, continued to collaborate with Alfred Hitchcock. He wrote such memorable Hitchcock films as *The 39 Steps* (1935) and *Sabotage* (1936). These opportunities led him to a successful screenwriting career, working with other esteemed directors like Cecil B. DeMille and earning him an Academy Award nomination for the script of Hitchcock's spy-thriller *Foreign Correspondent* (1940).

The topics of sexual violence and self-defense still resonate today, as the play resurfaced in London at the Mercury Theatre in 2022. The modern version, enhanced by Mark Ravenhill, touched on other pertinent social issues, like the distrust of police. A reviewer for *The Guardian* pointed out that for the 2022 *Blackmail*, it's not just Bennett's play but also Hitchcock's cinematic version that casts a shadow: "One smart design touch among many is the silhouette of a ghostly nightdress hanging in a window, evoking the image of Anny Ondra in the film."[8]

While most critics praised the mixture of sound and silence in Hitchcock's film, we wanted to get a modern-day perspective on film criticism. We spoke to acclaimed critic Paul McGuire Grimes, who gave us an insight into how he views and reviews films:

Meg: Tell our readers about your background and how you became a film and television critic.

Paul McGuire Grimes: I grew up loving the movies and watched *At the Movies with Siskel & Ebert* religiously as a kid, taking notes in my notebook about which way their thumbs went. I never thought it would be a reality to be a film or television critic, but I knew I always wanted to be involved in the movies. I had a passion for acting, so I got a BFA in Music Theater from Viterbo University. Cut to many years later when I wasn't working as an actor, so I started a movie review blog to share my love of movies. One thing led to another, and I reached out to KSTP about being the film critic on their daily lifestyle show *Twin Cities Live*. I did a screen test and have been featured on their show since 2014. I have since been featured on multiple radio stations like My Talk 107.1 and WCCO Radio, have interviewed A-list Hollywood actors and directors, and am a member of many critics guilds and am a Rotten Tomatoes–approved critic.

Kelly: What was your first experience with Hitchcock? Do you have a memory of seeing one of his films for the first time?

Paul McGuire Grimes: I don't quite remember the first time I watched a Hitchcock film, but it's safe to assume it was *Psycho*. That tends to be the gateway entry into appreciating Hitchcock's films. It certainly is the most popular, dissected, quoted, referenced, or spoofed. *Psycho* is frequently featured on Scariest Movies of All Time lists thanks to the infamous shower scene and Bernard Herrmann's shrieking violin score. My memories of watching Hitchcock films come later in life. The first time I saw *The Birds* on the big screen was for some anniversary screening, and the teens sitting next to me just laughed the entire time. Yes, Hitchcock had a wicked sense of humor, but I don't believe these teens were laughing with Hitch; more so, they were laughing at the film. A better memory comes with seeing his films with my mom as part of the Minneapolis Hitchcock Festival. It's a yearly series presented by the Trylon in partnership with the Riverview Theater and Heights Theater. My mom and I have attended many of these screenings, and it's that type of bond and lasting memories that the movies can have on the audience.

Meg: What is your take on films like Hitchcock's and how they are watched and critiqued today? Do you think it's important we watch these works with a modern eye?

Paul McGuire Grimes: Alfred Hitchcock is a good example of someone who in today's culture would be cancelled and dismissed by the industry. I bristle at even using those terms, but his treatment of his leading ladies, the Hitchcock Blondes, is widely known to be nefarious and inappropriate. There are two angles here. You can certainly view these films with a modern eye in terms of behind-the-scenes allegations and learning the history of how he treated his stars. The other angle is viewing the actual films with a modern eye to how Hitchcock executed the stories. He's not called the master of suspense for nothing. He had a keen visual eye and would storyboard everything. He frequently had the whole movie mapped out in his head based on the original source material. One good example of this is for *Rear Window*. John Michael Hayes wrote the dialogue, but Hitchcock worked with him to flesh out the other elements of the screenplay based on the images he already had in place of what the sets and camera movements looked like. Hitchcock leaves clues in the images on screen and trusts the audience to put the pieces together. You could watch his films on silent and understand everything that's going on based on the images alone.

Kelly: We agree!

Paul McGuire Grimes: It's also fascinating to watch some of Hitchcock's films featuring LGBTQ writers or themes. These films came at a time during the Hays Code (1934–1964) where there were strict guidelines on how sexuality or sexual persuasions could be presented on screen. Movies like *Rebecca*, *Rope*, and *Strangers on a Train* have been reanalyzed in recent years, specifically looking at the sexuality of its characters and how Hitchcock would leave subtle hints noting their sexuality without specifically stating it.

Meg: It's so true! It makes us wonder how many audience members caught it on first viewing when the films were first released.

Paul McGuire Grimes: Modern audiences have a different idea and need when it comes to horror and suspense. Take a look at slashers, torture porn, and found-footage subgenres of horror. It seems like modern audiences thrive on shocking images, jump scares, copious amounts of blood, and torture with little character development needed. There's no way that 2023's *The Exorcist: Believer* could live up to the original film released fifty years ago, as filmmaking and audiences have changed. Even trying to replicate Hitchcock rarely works. Gus Van Sant's shot by shot remake of *Psycho* (1998) flopped big time. Only Hitchcock can do Hitchcock.

Kelly: As a critic, how do you go into watching a film for the first time? Do you do any research ahead of time or go in cold? Do you ever rewatch a piece of media before forming your final opinion?

Paul McGuire Grimes: I like going into a film cold. Some of the best moviegoing experiences I have are when I'm taking in a film without watching a trailer, without reading a review or seeing behind-the-scenes footage. I can experience a film with little to no preconceived notion of what I think it should be. I rarely watch trailers anymore, as they tend to give away too much or can be misleading.

Meg: We love going into a movie cold too!

Paul McGuire Grimes: I dive into my research after I've watched the movie to learn more about filmmaker intentions and how he worked with his cast and crew. You can learn a lot about storytelling by how they work with their cinematographers and the story they're trying to tell with the images conveyed on screen. My reviews are based on first-time viewings. I love watching movies or television series multiple times to see how my opinion on them changes with different viewings. There are movies that I haven't liked as much on second viewing and vice versa. Sometimes I appreciate a movie more on second viewing, as I go in with different expectations.

Kelly: That's initially how our podcast, *Horror Rewind*, came about! We rewatched movies from our childhood to see if they lived up through the lens of today. We found that nostalgia played a big role in our memories of some of our favorite movies[9].

Thank you to Paul McGuire Grimes for sharing his process of how he reviews films! He has inspired us to continue to avoid trailers and spoilers before viewing a film for the first time and to go in with an open mind.

CHAPTER THREE

The 39 Steps

When England, and the entire world, was entrenched in the realities of World War I, Alfred Hitchcock was only a schoolboy of fifteen. His father died in late 1914 at only fifty-two, altering Hitchcock's home life, while the devastation of war raged outside. He once told an interviewer that "the first time he experienced genuine fear—as opposed to the enjoyment of fear—came when enemy bombs dropped on London. He was home with other family members, and they all fell to the floor. His mother took refuge under a table and cowered there, murmuring prayers."[1]

Naturally, the war left an impression on the youthful Hitchcock, who was rejected from the military. There is speculation that it was his already burgeoning weight that was the culprit of his dismissal, while others think it could be his glandular condition, and not the weight itself, that made him unable to be a soldier. Whatever the cause, Hitchcock didn't let it stand in the way of his patriotism, as he joined the Royal Engineers in 1917 as a cadet. As a part of the Royal Engineers, Hitchcock would've participated in local marches, along with military drills.

> In 2024 the top reasons for disqualification from the military include illegal drug use, alcohol dependence, not meeting height and or weight requirements, and certain law violations.[2]

The 39 Steps is a wartime film, steeped in the intrigue of international espionage. Based on the 1915 highly popular novel by John Buchan, *The 39 Steps* has been adapted for several films, TV shows, radio plays, and theater productions. There was even recent rumor it would be brought to Netflix starring *Sherlock* (2010–2017) and *Doctor Strange* (2016) star Benedict Cumberbatch.

The man at the center of the intrigue, Richard Hannay, goes on to be featured in four more novels by John Buchan, in one of the first examples of the man-on-the-run archetype. This prolonged suspense of Hannay being wanted for murder as he travels the globe is, of course, perfect for Alfred Hitchcock's skill in bringing his audience to the edge of its seats. The film differs from the novel but still maintains the patriotism of Buchan's work, as Hannay risks life and limb for his beloved England.

Unlike the lesser-known actors in *Blackmail*, stars of *The 39 Steps*, Robert Donat and Madeleine Carroll, had been in Hollywood films prior to their work with Hitchcock. British charmer Donat had just come off a major role in *The Private Life of Henry the VIII* (1933) and was in demand in the United States. After his turn as Richard Hannay, he would win the Academy Award for Best Actor for *Goodbye, Mr. Chips* in 1939. Yep, you read that right; he beat out Clark Gable in *Gone with the Wind* for that golden Oscar!

Like her costar, Madeleine Carroll's star was rising, and with the help of *The 39 Steps*'s success she went on to be the first British actress to be offered a major American movie deal. Thanks to her deal at Paramount, Carroll was a leading lady with costars like Gary Cooper and Bob Hope.

The 39 Steps was an instant hit in Europe and across the pond. Even esteemed director Orson Welles was known to be a fan of the film, and it did well with average moviegoers too. A critic for the *New York Times*, Andre Sennwald, wrote at the film's release, "There is a subtle feeling of menace on the screen all the time in Hitchcock's low-slung, angled use of the camera. But the participants, both Hannay and his pursuers, move with a repressed excitement that adds significance to every detail of their behavior."[3]

Many innovators are not appreciated in their time, but Alfred Hitchcock was impressing fellow directors, critics, and audiences alike. He continued his dominance in the field of spy films with the subsequent *Secret Agent* (1936), starring the familiar Madeleine Carroll. Next came British espionage thriller *Sabotage* (1936), in which a bomber has set his sights on London. This lifelong interest in spy thrillers would show itself over the next few decades with Hitchcock's later films like *Notorious* (1946), *Torn Curtain* (1966), and, what some consider his masterpiece, 1959's *North by Northwest*.

One of our most valued resources for researching Alfred Hitchcock is *Hitchcock/Truffaut*, a book by fellow revered filmmaker François Truffaut. Written in interview style, the book was first published in Truffaut's native French in 1966. It's a verbatim account of the two legendary filmmakers' 1962 discussion of Hitchcock's work.

François Truffaut (1932–1984) is known for his contribution to the French New Wave. "The explosion of creative innovation that emanated from France in the late 1950s and early '60s forever altered the course of film history by opening up new avenues of stylistic experimentation and trumpeting the concept of the 'auteur' director, whose aesthetic vision and thematic obsessions took center stage."[4]

Truffaut is best known for his film *The 400 Blows* (1959), which we both watched in our college film classes, as it is considered, perhaps, the most indelible example of the French New Wave phenomenon. Although he was not French, Alfred Hitchcock was an auteur becoming popular in the same era, thus heavily influencing Truffaut. In his conversational book, Truffaut asks Hitchcock numerous questions about filmmaking, with an eagerness we know we'd feel if given the chance. The following is an excerpt in which Truffaut points out Hitchcock's style emerging in *The 39 Steps*:

Francois Truffaut: Incidentally, on re-seeing your version of *The Thirty-Nine Steps*, I realized that it's approximately at this period that you began to take liberties with the scenarios, that is to attach less importance to the credibility of the plot, or at any rate, whatever necessary to sacrifice plausibility in favor of pure emotion.

Alfred Hitchcock: That's right!

Francois Truffaut: For instance, when Robert Donat is leaving London, on the train, he runs into a series of disturbing incidents. At any rate, that's the way he interprets what he sees. He thinks the two persons sitting opposite him in the train compartment are watching him from behind their papers. And when the train stops at a station, through the window, we see a policeman, standing at attention and staring straight at the camera. There are indications of danger everywhere; everything is seen as a threat. The deliberate buildup of the mood was a step in the direction of American stylization.

Alfred Hitchcock: Yes, this was a period when there was greater attention to detail than in the past. Whenever I embarked on a new episode, I would say to myself, "the tapestry must be filled here" or "we must fill out the tapestry there." What I like in *The Thirty-Nine Steps* are the swift transitions. Robert Donat decides to go to the police to tell them that the man with the missing finger tried to kill him and how the Bible saved his life, but they don't believe him, and suddenly he finds himself in handcuffs. How will he get out of them? The camera moves across the street, and we see Donat, still handcuffed through the window that is suddenly shattered to bits. A moment later he runs into a Salvation Army parade, and he falls in step. Next, he ducks into an alley that leads him straight into a conference hall. Someone says, "thank heavens, our speaker has arrived," and he is hustled onto a platform where he has to improvise an election speech. . . The rapidity of those transitions heightens the excitement. It takes a lot of work to get that kind of effect, but it's well worth the effort. You use one idea after another and eliminate anything that interferes with the swift pace.[5]

If you're interested in reading more of *Hitchcock/Truffaut,* we suggest you search in thrift shops and online. Meg's 1983 copy is packed full of behind-the-scenes photos of the master at work.

With all films, we want to make sure to point out Hitchcock's cameo, and we have to admit that in *The 39 Steps* we missed it! Thankfully the internet pointed us in the right direction, and we were able to spot both Hitchcock and screenwriter Charles Bennett in line for a bus only a few minutes into the film. Hitchcock even throws a bit of trash into a bin. So far, this is the most subtle of the director's appearances—how apropos for a film about spies!

The character Mr. Memory (Wylie Watson) commits fifty new facts to memory every day in *The 39 Steps*. How does the human memory work, and is this feat possible? The process for remembering things starts with encoding, or how information is learned. We take in information and understand it according to our senses. Next, we store the information in either our long-term or short-term memory. Our long-term memory can hold a lot of information, but it's up to the final step, recall, to prove how much we can remember. How can we improve our memory? According to the Mayo Clinic, everyday exercise, good sleep, organization, brain games, and eating healthy can improve memory.[6] Can you learn fifty new facts a day? Maybe. Because we're all individuals, we learn at various paces and retain information differently. Some tips are to repeat key concepts, write things down, and even practice saying things out loud that you'd like to remember.

Mr. Memory reminds us of one of our favorite characters, Sherlock Holmes, and his "mind palace" in the BBC series *Sherlock* (2010–2014). In it, he employs techniques to store and retrieve memories. If you'd like to experiment with this form of memory building, it's recommended that you choose a place that you know well, like your home or office, then plan a route through it. Next, create a list of things you want to remember and place them throughout the building. Exaggerating the images in your mind will help you conjure them when it's time to put this technique to use.[7]

It's announced at the magic show that "Mr. Memory has left his brain to the British Museum." How long have people been donating their bodies for scientific purposes? In our research, we discovered that bodies have been donated for anatomical studies since prehistoric times and that it was regarded as a noble deed. Between 1514 and 1564, Andreas Vesalius, considered the father of modern anatomy, conducted the first scientifically established human dissections for learning medicine through anatomy. As a result of his work, medical schools of dissection began to be established, and this in turn led to an increased demand for cadavers in Europe. Because of the growth of unethical practices to procure bodies (like grave robbing and even murder!), laws were created, and donations became regulated.[8]

The men start to fight at the show in *The 39 Steps*. What is this masculine urge to throw fists? Johnathan Gottschall, author of *The Professor in the Cage: Why Men Fight and Why We Like to Watch*, wrote that "a diverse array of species—from beetles to birds to bears to mantis shrimp—all share strikingly similar 'dueling' behaviors."[9] Men may feel that they have a duty to fight to defend their own reputations and their loved ones' honor. It's recommended that if you feel like fighting, take a deep breath, remove yourself from the situation, and try to remain calm. Feeling anger isn't negative; it's how you handle the emotion that can be a problem.

Those in charge in *The 39 Steps* ask the orchestra to "play something to stop the panic." It feels like the scene in *Titanic* (1997) when the ship is going down, and the band is told to play upbeat music for the passengers. Does music affect our mood? Absolutely! Music can enhance our mood or change it entirely; it can relieve stress, help with our memory, and even encourage brand recognition when used in ads.[10] From our own experience, we can vouch for music affecting our moods, pumping us up for important presentations, and helping us relax after a long, busy day.

This movie features a female agent, perhaps unexpected at the time by audiences. What is the history of females as spies? Harriet Tubman not only led over three hundred enslaved people to freedom during the Civil War but was also a spy who collected intel for the Union. She is also credited with rescuing over 750 enslaved people on a raid in which she led three gunboats. During World War I, Mata Hari, a dancer turned spy, convinced military officers and diplomats to share secrets with her to help the Germans. She was caught and sentenced to death. During World War II, entertainer Josephine Baker was able to slip past enemy lines due to her star power and delivered secret messages written in invisible ink. Women are even more active today in organizations like the CIA.[11]

A hymnbook in the pocket of Hannay's overcoat stopped a bullet from piercing his heart in the film. Is this possible? Yes, depending on the speed and size of the bullet and the thickness of the book. An experiment was done in 2018 that showed a handgun bullet was stopped by the thickness of two books at close range while three books stopped a bullet from an AR-15.[12] Depending on the thickness, many other household materials can be bulletproof, like brick, wood, Kevlar, and fiberglass.

The moors of Great Britain are featured in *The 39 Steps* as well as some of our favorite literature set in England, like *Wuthering Heights* (1847) by Emily Brontë and *The Hound of the Baskervilles* (1902) by Sir Arthur Conan Doyle. Why do they hold such allure? The damp, foggy, bleak topography has fascinated and terrified people for decades. The contemporary definition of a moor is "an elevated, hilly, or mountainous region that is wild or, at most, only sporadically cultivated, but which may be managed through grazing, cutting, and burning."[13] Britain holds more of this type of land than any other nation in the world, at 15 percent of the total, and it feels like the perfect setting for a mystery, romance, or thriller.

Hannay and Pamela, played by Madeleine Carroll, are handcuffed together, and this puts them in some interesting predicaments. Hannay has Pamela sign the guestbook at the hotel because he can't write with his left hand. What is the science behind right-handed versus left-handed people? According to *Medline Plus*, "In Western countries, 85 to 90% of people are right-handed and 10 to 15% of people are left-handed. Mixed-handedness (preferring different hands for different tasks) and ambidextrousness (the ability to perform tasks equally well with either hand) are uncommon."[14] While in the current era we determine our handedness in early childhood, previous generations were taught to write with their right hands even if they were decidedly left-handed. Recent studies have shown that up to forty genes contribute to the trait of which hand you will be most comfortable using the most.[15]

In *The 39 Steps*, Hannay mentions Pamela's skirt is soaked and she might get pneumonia. Is there any truth to this old adage? Not really. Having wet hair or clothes won't cause you to catch a virus, but if you get hypothermia or become overly cold, your immune system may become weaker. This, in turn, will make you more susceptible to catching an illness. Experts recommend going out in colder weather for activities and exercise because it's less likely for you to spread germs outdoors.[16]

It seems as if Pamela may be suffering from Stockholm syndrome with Hannay. This condition, which got its name after a 1973 bank robbery in Sweden, refers to when hostages begin to share feelings or a bond with their captor. The inverse of this is called the Lima syndrome, when abductors develop sympathy for their hostages. This condition got its name in 1996 after an abduction in Lima, Peru.[17] The duo in this film may be experiencing both!

Photographic memory is featured in *The 39 Steps*. Is this humanly possible? I (Kelly) used to study for quizzes and remember the pictures that were on the textbook pages that contained the concepts I was defining. Although this isn't photographic memory, it did allow me to remember things to use later. People with exceptional memory are said to have highly superior autobiographical memory (HSAM); fewer than one hundred people worldwide have been identified with having this condition.[18] People with HSAM are able to remember exact dates and recall things in great detail. For those of us who don't have this talent, it's important to use multiple senses to remember something. Listen, look, and repeat key facts you'd like to remember.

CHAPTER FOUR
The Lady Vanishes

Like many of Hitchcock's films, *The Lady Vanishes* is based on a British novel. *The Wheel Spins* (1936) by Ethel Lina White was a popular mystery. Several of White's novels were adapted to film and theater, especially after the success of *The Lady Vanishes.* Considered as popular as Agatha Christie in the United Kingdom in the 1930s and 1940s, White's work hasn't endured with the same staying power as Christie's. There has been a resurgence of interest in Ethel Lina White, however, including a 2013 BBC production of *The Lady Vanishes*, which stayed closer to the source material's plot than Hitchcock's version.

> The premise of *The Lady Vanishes* was based on a legend of a woman who disappeared in the Palace Hotel in Paris in the 1880s. The story goes that the woman had the bubonic plague and the hotel hid the fact so Paris wouldn't have to evacuate.[1]

There is also a 1979 *The Lady Vanishes,* a more direct remake of Hitchcock's version, starring Cybil Shepard, Elliot Gould, and Angela Lansbury. It's notable that this was the last film produced by Hammer Films (known for its horror movies of the 1950s) before a nearly thirty-year hiatus.

After a few box-office disappointments, *The Lady Vanishes* cemented Hitchcock's status as a successful director, one who could make it big in America. Due to *The Lady Vanishes* being a worldwide hit, Alfred Hitchcock was finally able to move to the land of glitz and glamor:

Hollywood. Thus, *The Lady Vanishes* is considered the last film in Hitchcock's British era.

One such believer in Hitchcock was the producer David O. Selznick, known as the producer of *Gone with the Wind* (1939) and one of our favorite Hitchcock films, *Rebecca* (1940). Once Hitchcock moved to California, he not only had the benefit of a sunnier climate but also was afforded bigger budgets and well-outfitted studios, something he didn't have while making *The Lady Vanishes* at Islington Studios. Because of an economic slump in Britain in 1937, the studio wasn't in its best shape. In *Alfred Hitchcock: A Life in Darkness and Light*, Patrick McGilligan explained the ingenuity needed to make films there:

> "*The Lady Vanishes* would be a reunion with cameraman Jack Cox, and one last Hitchcock film produced at Islington. Cox had to summon his old resourcefulness on the once grand soundstages now small and threadbare. . . . The pictures shot at Islington during this era were conveniently set in train compartments, light houses, and cramped prison cells. Alex Vetchinsky, the resident designer, had become an expert at that sort of inexpensive single set, and his influence is conspicuous on *The Lady Vanishes*. Hitchcock was encouraged to make abundant use of his full arsenal of back projection, trick shots, and miniatures to maintain the illusion of a real train in constant motion."[2]

To a modern audience, the use of miniatures is quite obvious from the beginning of *The Lady Vanishes*, in which we are introduced to a snow-dappled Swiss-like town. A vehicle drives by that we can tell is no bigger than a Matchbox car. The use of miniatures in film is an oft-used device that can help with both showing the impossible and helping with the film's budget. In his groundbreaking 1902 film *A Trip to the Moon*, Georges Méliès made use of miniatures to show a rocket traveling to— yep, you guessed it—the moon. *King Kong*, one of the earliest monster films, marked another landmark in the use of miniatures. Directed by Merian C. Cooper and Ernest B. Schoedsack, the film was released in 1933. The scene where King Kong fights with the snakelike dinosaur is one of the most famous special-effect moments in the film. The scene was accomplished through stop-motion animation, matte paintings, real water, smoke, foreground rocks with bubbling mud, a miniature set, and two miniature rear-screen projections of the characters Driscoll and Ann. For the film, four models of King Kong were built, two jointed eighteen-inch aluminum, foam rubber, latex, and rabbit-fur models; one jointed twenty-four-inch model; and a small model of lead and fur. The dinosaurs were made in the same fashion as King Kong, and in some models football bladders were used to simulate breathing. The Venture Railway cars and the warplanes were all miniature models.[3]

Hitchcock was known to use miniatures throughout his career, including in *Rope* (1948), a film known for its feat of lengthy takes. Shot almost entirely in a set made to look like an apartment, *Rope* is a technical marvel in which every camera angle and actor had to be in rhythm in order not to break a long take. This included the background seen through the windows. While it seems like a small detail when the action is happening inside, Hitchcock spared no expense or expertise to make the miniature city believable:

> The miniature reproduced about thirty-five square miles of the Manhattan skyline, taking in such familiar landmarks as the Empire State Building, St. Patrick's Church, the Hotel Astor, Radio City, and the Chrysler and Woolworth buildings. The 12,000-square-foot cyclorama backing was three times as wide as the apartment and was laid out in a semi-circle so the camera could be moved around freely without compromising the background. The buildings, built-in forced perspective, were similarly arranged. The closer structures were three-dimensional and equipped with steam pipes inside to feed "smoke" through chimneys and stacks. It was noted initially that the steam rose too rapidly and too high for the scale of the set. Dry ice, placed over the pipes, solved the problem. The chilled steam rose lazily and drifted away.[4]

We wonder what happened to that miniature Manhattan. We'd pay a pretty penny to see such a fascinating piece of film history. If you want to learn more about *Rope*'s inspiration from real-life murderers, read our 2021 book *The Science of Serial Killers*.

> Many movies have used long takes, like Hitchcock used in *Rope*, including John Woo's *Hard Boiled* (1992), *Birdman or (The Unexpected Virtue of Ignorance)* (2014), and the incredible trench-run scene in *1917* (2019).

Espionage is still Hitchcock's main concern in *The Lady Vanishes,* beginning in the fictional European country of Bandrika. Interestingly, a movie studio has proudly borrowed the fake country's name. Bandrika Studios has made such eclectic fare as the FX series *Fosse/Verdon* (2019) and the film *The House with a Clock in Its Walls* (2018). The country itself has a Swiss look to it, with a language mash-up of several different European dialects. Bandrika joins an illustrious list of fictional European countries like Andalasia (*Enchanted,* 2007) and of course the gothic cartoon of our youth took that place in Bratislava (*Count Duckula,* 1988–1993.)

Because of the issues with space in Islington Studios, most of the film takes place on a train heading to England. When an innocent old lady Miss Froy (May Whitty) goes missing, it is only Iris Henderson (Margaret Lockwood) who seems to remember she was ever on the train in the first place. With the help of handsome Gilbert (Michael Redgrave), Iris ends up stumbling into a conspiracy of murder of and, yes, espionage. The seemingly nondescript Miss Froy is actually a British spy! She has a secret musical tune that she must convey to London officials before she is murdered by enemies on the train. While the screenplay was written by Sydney Gilliat and Frank Launder, Hitchcock tweaked their ending to show that Miss Froy lives through the gunfire and indeed saves the day. It's a memorable, cheeky ending for a film with such serious subject matter as murder. It is this balance of light and dark that truly makes a Hitchcockian film. In his article for *Medium*, film journalist and producer Jeffrey Michael Bays argues that comedy is at the heart of all of Hitchcock's films:

You might be surprised when I say Alfred Hitchcock was a master of film comedy. After all, this is the same director who gave us the grueling oven murder in *Torn Curtain* (1966).

But Hitchcock did say, "every film I make is a comedy," and even considered *Psycho* (1960) a practical joke. And the more you watch his films, it becomes apparent that even the deadliest situations have an undercurrent of facetious wit, using comedy to actually heighten the suspense. The key is that Hitchcock's humor is specifically aimed at the audience. It's a part of the game he activates, turning us into active viewers, making us aware of this master storyteller pulling the strings behind the camera. When we laugh, we're laughing at his prankster omnipresence felt during the film.[5]

Bays continues his article with many examples of quirky characters and non sequiturs that make his movies both funny and frightful. He maintains that this balance of the two is what Hitchcock was a true master of. And we certainly agree.

Don't worry—when it comes to his comedic commitment to appearing as an extra in his films, Alfred Hitchcock won't disappoint you. He, of course, has a cameo in *The Lady Vanishes*. Near the end of the film, he walks by with suitcase in hand and a cigarette hanging from his mouth. Like in *The 39 Steps*, it's a blink-and-you-miss-it moment! We recommend, if you're having trouble with the Where's Waldo nature of his cameos, that you check out the website The Hitchcock Zone (https://the.hitchcock.zone), which has handy screenshots.

Although Hitchcock never technically cast himself in a role, we became curious about the method of choosing actors for a project. We spoke to Kate Caldwell, a casting director based in the United Kingdom, about her process for this important part of filmmaking:

Meg: Tell us about your career in casting and how you got your start.

Kate Caldwell: I have always loved TV and movies. As a fat kid growing up in Fargo, I much preferred sitting in front of the television on a Saturday or in the darkness of a movie theatre than playing outside. I really wanted to be an actor, but I didn't have the courage to even try out for plays.

Kelly: I was extremely shy as a kid too!

Kate Caldwell: I went to Moorhead State University to become a journalist. My goal was to work for a big movie magazine like *Premiere* or *Movieline*—neither of which even exist anymore. I think I was in my sophomore year when I happened upon Intro to Film taught by Ted Larson, and I thought that sounded fun. I would find out later that a lot of students took film classes looking for an easy A, and many would end up dropping out because Ted Larson did *not* do easy. After that first class, I was hooked. I started taking every film class I could fit into my schedule until one day my Mass Communications advisor informed me that I had a Speech Communications major thanks to all of these film and TV classes I had been taking, so I graduated with a Speech Communications degree and was one Econ class short of a Mass Comm degree as well.

Kelly: We had a similar path!

Kate Caldwell: I loved studying film classics and still love watching them to this day. These were auteurs who created each genre from the ground up. The films had a different look and feel to them. They weren't overproduced; they were gritty and real. After I graduated, I wasn't sure what I was going to do next. I hated prying into people's private lives, so I knew I was never going to make it as an entertainment journalist. (TMZ is my ultimate nightmare.) I don't really remember this, but everybody I went to college with said that I always talked about how someday I was going to move to L.A. I asked my best friend, Tamara, if I ever said, "What the hell, I [am] going to go there," and she said, "Not really; you just always said you would end up there." In 1996 my friend Danny mentioned that another MSU alum, Doreen, was moving out of her apartment in L.A., and her roommate needed somebody to move in. "You always talk about moving there; here's your chance!" So, I did it. It was completely out of nowhere, and I think I figured I would go out there, check it out, and then end up back in Fargo where I'd live the rest of my life saying, "I used to live in L.A." Doreen was a talent agent, and she gave me the opportunity to work on her desk for two weeks while her assistant was on vacation. I have nothing but respect for talent agents. They have to navigate the egos, insecurities, and needs of the actors as well as the producers, directors, casting directors, etc. I couldn't do it. They really earn their money. When Doreen asked me what I thought, I told her that it was the worst job that I ever had, and it was not for me. When I told her I wasn't sure what I did want to do, she asked, "What do you like? What do you know?" And I immediately answered, "Actors! I know actors!" And I really did. I didn't just know stars; I knew character actors. I knew those actors who, when you pass them in the street, the average person thinks they went to school with them, but I could tell you every TV show they guest starred on and every movie they had three lines in. She said, "Great! Let's get you an internship!"

Meg: That's incredible!

Kate Caldwell: She called up Ulrich/Dawson/Kritzer Casting and asked if they needed someone to come work for them for free. Thankfully, the movie *Fargo* (1996) had come out, and I had the accent, so Eric Dawson said, "If you will do that accent for my wife anytime she calls here, you can work for us for free." And that's how I fell into my career.

Kelly: Anytime we tell people we're from Minnesota, they automatically go into the *Fargo* accent!

Kate Caldwell: Throughout the years, I was lucky enough to have a plethora of amazing bosses. I worked mostly in TV in the beginning, which was the best training ground imaginable. TV works fast so you learn how to multitask and get things done quickly. I eventually went to 20th Century Fox Feature Casting in order to break into movie casting. We were the studio, so we often oversaw casting of the movies, but every once in a while we got to do some boots-on-the-ground casting as well, which was fantastic. Eventually, I had my own show on DirecTV called *Kingdom* (2014–2017) and then partnered up with my BFF Melissa Kostenbauder, and we got nominated for an Emmy alongside Laura Rosenthal and Jodi Angstreich for a Netflix limited series called *Unbelievable* (2019). Shortly after we wrapped on that, I moved to Bristol, England, where I live with my husband, Rich. Suzanne Smith was kind enough to hire me to work with her and allows me to work remotely, although I go into the offices in London from time to time as well.

Kelly: What do you look for when casting a character?

Kate Caldwell: I always visualize the character as I'm reading it and make lists of the actors that come to me. Sometimes they are pie-in-the-sky ideas, but they're really just to get me going creatively. Eventually, you will have a conversation with the showrunner and director (if it's television) or the director (if it's a film), and that will often change the way you see things and change that list. The goal is to find what your creatives want but also keep a bit of you in there. It doesn't always have to be so on the nose. Your creatives may want a late twenties, handsome male for a role, but, if it makes sense, I will bring up something completely outside the box, like a late forties, charactery woman. Sometimes they go for it, sometimes they don't, but it never hurts to try. You want somebody that you know will bring life to a character but is also open creatively to what the director has in mind, someone who hopefully is a team player but has a strong sense of self so they can infuse the role with their own essence. And some people are just magic on screen. It's something you can't force; it's just something they have. It's a beautiful thing to witness.

Meg: What are your opinions on Alfred Hitchcock's casting choices?

Kate Caldwell: He really loved those beautiful blonde women, didn't he? But they were talented beautiful blonde women, so I really can't complain. . . . The ones that really stand out for me are Tippi Hedren in *The Birds*, Grace Kelly in *Rear Window* and, of course, Janet Leigh in *Psycho*. Stunning looking, every one of them, but also strong women with character flaws who brilliantly played strong women with character flaws. That's one of the things I always loved about Hitchcock movies—so many characters who aren't perfect, so many women with secrets to hide and who held their own against any man, things to like and dislike about them. Nobody was perfect, which made them all so relatable. His male choices were equally intriguing. Jimmy Stewart in *Rear Window* was sheer perfection. You need that everyman in that role—someone you can relate to and empathize with. That was Jimmy Stewart's bread and butter, and he was a master at it. And Joseph Cotten in *Shadow of a Doubt*—just wow, so handsome and charming. Joseph was one of my first crushes as a child. As I mentioned, I was a fat kid who loved to sit in front of the TV, which meant I caught a lot of old movies on random Saturday afternoons. I think I first saw him in *The Magnificent Ambersons* (1942), and I was hooked. Eventually, I caught him in a few *Alfred Hitchcock Presents* episodes, which led me to *Shadow of a Doubt*. Growing up Catholic, I knew I shouldn't be so attracted to a bad guy, but I couldn't help it because it was Joseph Cotten! Hitchcock clearly knew what he had in Joseph, which is why he cast him multiple times. He had great taste in character actors as well. Leo G. Carroll is a standout, as well as Clare Greet (arguably not so charactery, but a great actor).

Kelly: How do you think casting has changed from Hitchcock's era to current day?

Kate Caldwell: "The most obvious one is diversity. It took until recently for film and television to catch up with society and realize that there are multiple cultures and ethnicities in this world and all need to be represented. We haven't mastered this yet, but we are working on it. I'm hard-pressed to think of any diverse leading men or women in any Hitchcock movie. Canada Lee was in *Lifeboat*, but I wouldn't consider that a lead. The other is in the casting of more interesting faces instead of the general Western standard of beauty. The creative people involved are much more open these days to actors with more characterful faces in our leads. Yes, Margot Robbie and Idris Elba are always going to be highly sought after, but so are Barry Keoghan, Ke Huy Quan, Tilda Swinton, and Katy M. O'Brian, who are attractive in ways that are so different than the old chiseled-jaw-piercing-stare or buxom-blond-hair-blue-eyed standard of Hitchcock's time.

Meg: That's so true! Interesting and different-looking actors bring so much to the screen.

Kate Caldwell: Hitchcock often cast the beautiful people in the leads and the more charactery people as bad guys or secondary characters. We are finally learning to get away from that and to represent normality a bit more.

Meg: In regard to Hitchcock or your own experience, how do you view casting an unknown compared to a well-known actor or actress in a role?

Kate Caldwell: As a casting director, there is nothing more exciting than discovering new talent and helping put them in a role that introduces them to the world. Unknown actors are blank slates for an audience. They have no past roles for which most people may know them, nothing to muddy the waters. Stars come with baggage, through no fault of their own. You can't help but think about some of their past roles, especially when iconic. Sometimes people are able to get past that; sometimes they aren't. But with an unknown actor, they are essentially a blank canvas. Edward Norton in *Primal Fear* [1996] is the first one that comes to mind for me when I think of this. He was known in the theatre world but had done very little film at that point, so what he did on screen wasn't tainted by preconceived notions. It's a brilliant performance and one that Hitchcock would have admired greatly.

Thank you to Kate for sharing her story and expertise regarding casting! This has made us think about how changes in actors could really affect a script.

It's windy in the beginning of *The Lady Vanishes*. What causes wind? "Wind is caused by uneven heating of the earth's surface by the sun. Because the earth's surface is made up of different types of land and water, the earth absorbs the sun's heat at different rates."[6]

As we were watching the film, we noticed the height of people and wondered how this physicality has changed historically and globally over the years. I (Kelly) lived in the house my grandfather built for many years, and my husband, who is six feet, four inches tall, would always hit his head on the door frames! We attributed it to people being shorter one hundred years ago. Is this true? According to *Time* magazine, people are about two inches taller now than we were one hundred years ago.[7]

A tune must be remembered to pass along a secret spy message in the film. How accurate is human memory when it comes to musical tones? Some people are phenomenally adept at remembering musical phrases. This skill is known as musical memory and may be related to the phonological loop. According to Diana Deutsch from the University of California, "We have argued for the view that music is represented in the mind of the listener in the form of coherent patterns that are linked together so as to form hierarchical structures. We have also examined the system underlying memory for tones and have explored a number of paradoxical illusions together with their implications. The system that we are dealing with is very complex, but an understanding of its characteristics is slowly emerging."[8] Sometimes, musical tunes get stuck in our heads even when we don't want them to. The phenomenon is known as an earworm, and as many as 98 percent of the Western population has experienced this. Earworms are most often linked to subconscious memory, emotions, and even anxiety.[9]

Folk dancing is shown in *The Lady Vanishes*. What is the history and science of why we dance? Research shows that dance is important in social interaction and can influence mate preference.[10] Because dancing includes our senses, we can easily judge someone by how they look, smell, and sound while moving to music. No pressure the next time you're on the dance floor! What other animals dance? Some birds, including parrots and cockatoos, are known for their dance-like moves and have even been shown to groove to a beat. Asian elephants have also been recorded dancing.[11] If a neurobiologist ever comes knocking to assess our dance moves, we too may vanish like the lady in the title of this film. Until then, you'll find us on the dance floor.

CHAPTER FIVE

Rebecca

When we're interviewed for podcasts or blogs, we field quite a few questions about our favorite books. Since the start of her career, Meg has maintained that Daphne du Maurier's 1938 novel *Rebecca* had the most profound effect in inspiring her to write and research gothic fiction. Considered by many to be a modern retelling of Charlotte Brontë's *Jane Eyre* (1847), *Rebecca* has been a vital piece of suspense entertainment throughout the decades from Hitchcock's adaptation in 1940 to the Netflix version in 2020 starring Lily James. The novel was immensely popular in its time and has been cited as an inspiration for contemporary authors (not just Meg!) like Stephen King. *Rebecca* is lauded for its potent atmosphere, a gothic throwback that makes the unnamed main character feel as if the ghost of her husband's former wife is haunting her in every room of the mansion on the sea, Manderley.

> Critic Kate Kellaway wrote, "Du Maurier was mistress of calculated irresolution. She did not want to put her readers' minds at rest. She wanted her riddles to persist. She wanted the novels to continue to haunt us beyond their endings."[1]

Dripping with suspense and thrills, *Rebecca* was basically *made* to be directed by Alfred Hitchcock. Producer David O. Selznick saw this as a match made in Hollywood heaven. As soon as Hitchcock settled in a new home in Los Angeles, he signed a contract with Selznick International. In 1939, right before *Rebecca*, he directed another du Maurier novel,

Jamaica Inn (1936), which, like the novel itself, didn't have the staying power of *Rebecca*. According to Hitchcock biographer Patrick McGilligan, Hitchcock only directed the film because he was a fan of actor Charles Laughton: "Hitchcock didn't care a fig about *Jamaica Inn*; he had agreed to direct it largely out of desperation."[2] Apparently, the director wasn't thrilled by the final product either. This led to him making certain he had more control over the making of *Rebecca*. That being said, Hitchcock, too, finds issue with his film version of *Rebecca*. In *Hitchcock/Truffaut*, François Truffaut asked if he was satisfied with it, to which Hitchcock replied, "Well, it's not a Hitchcock picture; it's a novelette, really. The story is old-fashioned; there was a whole school of feminine literature at the period, and though I'm not against it, the fact is that the story is lacking humor."[3]

Okay, so maybe *Rebecca* doesn't have Hitchcock's signature dark humor, but he nailed the suspense. When Meg got the VHS copy of the film as a teen, it quickly became one of her favorite Hitchcock films. He captured the eerie feel of Manderley, the innocence of the second Mrs. de Winter (Joan Fontaine), as well as the stifling feeling that Rebecca is everywhere. Don't just take Meg's opinion; it also won Best Picture at the Academy Awards. Film critic and actor Jay Jacobson explained what makes *Rebecca* so special on his blog, *Jay's Classic Movie Blog*:

> For a film with not a lot of action, Hitchcock keeps the pace of *Rebecca* moving at light speed. Tension never lets up as we watch the young bride discover things are amiss, while she nervously lives in the shadow of Rebecca. Hitchcock knows just what to show and how to show it to get the most tantalizing and intense effect possible. One example of how he can heighten the most mundane dialogue, is the scene in which Maxim [Laurence Olivier] explains Rebecca's final night to [the] main character. Instead of choosing to focus on Maxim or the main character, or showing the typical flashback, Hitchcock uses the camera to illustrate Rebecca's movements. In accordance with Maxim's description of her actions, the camera shows the empty sofa where she sat, rises where she stood, and moves as if following her where she walked. Not only

does this force us to vividly picture Rebecca in our minds, but it also reinforces her eerie, ghostlike presence.[4]

One of the reasons Rebecca may have not resonated as strongly with Hitchcock himself is because of his complicated relationship with producer Selznick. Alright, *complicated* may be misleading, as numerous books and articles have been written on their heated relationship. In fact, it was so bad that Hitchcock only made two more films with Selznick after signing his sizable contract with Selznick International, *Suspicion* (1941) and *The Paradine Case* (1947). It is rumored that this rift between the two titans of Hollywood was because they were both hard-headed about being in control of their projects. While it makes for poor partnerships, this auteur sensibility is what makes Hitchcock's body of work so impressive.

Literature professor Dennis Perry presented a lecture to his students about the fight behind the scenes of *Rebecca* between two men with two distinct visions of the film. Journalism student Eric Baker covered the lecture:

> On one side, many argue that Selznick was the real force behind the film. Coming off the success of *Gone with the Wind* [1939] and described by Perry as "large and in charge," Selznick may have been just a producer, but he prided himself on his intimate involvement with the making of each of his films. When Hitchcock's team submitted their original draft of the screenplay, Selznick replied with a three-thousand-word response that alleged they had "removed all the subtleties" and demanded changes. Selznick got his wish but still provided a ten-page list of suggestions to the revised draft. . . . But, despite Selznick's ever-present role in production, Hitchcock was not one to be intimidated into altering his own artistic vision. Hitchcock was far less concerned with staying true to his source material—a major concern for Selznick—and was more focused on what Perry described as "turning good books into great cinema." This, of course, did not sit well with Selznick and was a point of contention between the two. Hitchcock undermined

Selznick through his signature in-camera editing technique where he would only shoot the required scenes without providing extra footage for later editing.[5]

Whereas Selznick and Hitchcock were perhaps too much alike to work together, another partnership was far more important to the director. Married since 1926, Alma Reville and Alfred Hitchcock would spend their lives together until his death in 1980. She died a mere two years later. Reville was only one day younger than her husband and had grown up around the film industry, as her father worked at Twickenham Film Studios in London. Hitchcock and Reville met when they both worked at the studio in Islington; she edited one of his first projects, a silent drama called *Woman to Woman*. So, she can say she was there from the beginning, not hovering in the background, but involved in her husband's work. She worked with other directors in the United Kingdom, too, but when she moved to the United States with her husband, she became his most valued partner in crime and suspense. She collaborated on scripts like *Suspicion* (1941) and *Saboteur* (1942) and watched early cuts of his films to make suggestions. When they first met at the studio, Reville was a superior to Hitchcock, so he kept his interest in her hidden for several years. Reville recounted their very first meeting when Hitchcock had been hired at the art department: "Newcomers who came into our world inevitably reacted with awe and bewilderment, but this one was different. He's strolled across the set with a deadpan expression, stopped to ask me where the production office was, and when I pointed to the building, he nonchalantly disappeared into it without saying another word."[6]

If you're interested in the Hitchcock-Reville union, we recommend you check out the 2012 film *Hitchcock*, starring Anthony Hopkins and Helen Mirren as the duo. It's an intriguing peek into what their marriage might have been like.

While the man himself could be a bit stoic, it was Hitchcock's ability to convey emotion through the camera that makes *Rebecca*—and, really, all his work—sing. In his 1940 review of *Rebecca* from the *New York Times*, Frank S. Nugent shares this belief (but not before he fat-shamed Alfred Hitchcock in the same article, calling him "round" and "fond of beef"!):

> So, *Rebecca*—to come to it finally—is an altogether brilliant film, haunting, suspenseful, handsome and handsomely played. Miss du Maurier's tale of the second mistress of Manderley, a simple and modest and self-effacing girl who seemed to have no chance against everyone's—even her husband's—memories of the first, tragically deceased Mrs. de Winter, was one that demanded a film treatment evocative of a menacing mood, fraught with all manner of hidden meaning, gaited to the pace of an executioner approaching the fatal block. That, as you need not be told, is Hitchcock's meat and brandy. In *Rebecca* his cameras murmur "Beware!" when a black spaniel raises his head and lowers it between his paws again; a smashed china cupid takes on all the dark significance of a blood-stained dagger; a closed-door taunts, mocks and terrifies; a monogrammed address book becomes as accusative as a district attorney.[7]

"Last night I dreamed of Manderley again," begins the film adaptation of *Rebecca*. How do most of us dream? Are dreams realistic, or do we float as the camera does in the film's opening scene? "Studies have revealed diverse types of dream content, but some typical characteristics of dreaming include it has a first-person perspective, it is involuntary, the content may be illogical or even incoherent, the content includes other people who interact with the dreamer and one another, it provokes strong emotions, and elements of waking life are incorporated into content. Although these features are not universal, they are found at least to some extent in most normal dreams."[8] While we all dream, we may not remember them when we wake up. To improve the memories of your dreams, think about them right when you wake up, and keep a dream journal.

Joan Fontaine's character is a paid companion. What is the history of this practice? Women in the eighteenth century until the mid-twentieth century couldn't take a position of work, so, instead, they became companions. This wasn't a servant role, but rather one in which the woman would spend time with her employer, help entertain guests, and have conversations with the lady of the house.[9] The position was created because upper-class women primarily stayed at home and employment opportunities for single women of a higher class were not abundant. Numerous examples of lady companions are in the works of Agatha Christie, Jane Austen, and Louisa May Alcott.

Mrs. Van Hopper (Florence Bates) talks about how Mr. de Winter has lost his wife. Mourning the death of loved ones is a difficult journey and looks different for everyone. What do experts recommend for going through the grieving process? First, it's important to know that grief and its timeline will look different for everyone. Don't ignore your feelings of grief; instead, confront them and work through them. Next, allow yourself time to grieve. It's more important to sit with your feelings than to pretend everything is okay. Finally, reach out and ask for help if you need it. If you know of someone who is grieving, it may be difficult to know what to say. But the important thing is that you make yourself available for whatever the person may need, like a meal, a shoulder to cry on, or a distraction. While watching *Rebecca*, we agreed that it would be nearly impossible to *not* think about the title character because the entire house was filled with her memories!

The new Mrs. de Winter must get used to the class dynamics at Manderley rather quickly and has some big shoes to fill from her predecessor. What are some tips to succeed in a new role? According to the *Harvard Business Review*, nearly half of all leadership transitions fail.[10] While this statistic applies to jobs, running Manderley fits as it is a leadership position. Experts recommend doing your homework ahead of time (know what you're getting into), be yourself, understand and manage relationships with staff and coworkers, have a positive outlook, and don't be afraid to ask for help.[11] There was no way the new Mrs. de Winter could have known what she was getting herself into, but these tips might have helped.

Everyone knew Rebecca except the new Mrs. de Winter. Is it a good idea to learn about our spouse's exes? Relationship experts say yes. Knowing why and how your partner's previous relationship ended could give you a clue as to what the future holds. It's also imperative to find out what the spouse's current rapport is with the ex and how he or she speaks about the person. Are they close? Are they enemies? Gauging the dynamic will only be helpful.

> The new Mrs. de Winter feels more intimidated by Maxim's previous wife than jealous of her. Studies show that people are more likely to be jealous of those who share similarities but are ahead in some way, whether it be wealth or status, that feels undeserved.[12]

It's revealed that Maxim never really loved Rebecca, and he may be partially responsible for her death, or at least her disappearance, but his name is cleared in the end, and Manderley is cleansed of Rebecca's memory by burning down.

CHAPTER SIX

Suspicion

We feel that *Suspicion* is a convincing cautionary tale of being charmed by the wrong man. And who better than to play the charmer than Hollywood icon Cary Grant? *Suspicion* marks the first collaboration of Hitchcock and Grant, though it won't be the last, or the most well-known. Grant was known for both his comedic and dramatic roles in movies as varied as the farcical murder comedy *Arsenic and Old Lace* (1944) to the tearjerker *An Affair to Remember* (1957), of which plot points are replicated in a more modern tearjerker, *Sleepless in Seattle* (1993). As Johnnie Aysgarth in *Suspicion*, Grant gets to play both fun-loving playboy and devious husband. This contrast of personality in one character is done deftly by Grant, only enhanced by Joan Fontaine's clever and, of course, suspicious, wife, Lina McLaidlaw. The Academy agreed, as it awarded Fontaine an Oscar for Best Actress as Lina. Whereas Hitchcock felt *Rebecca* (also starring Fontaine) was a bit too serious for his taste, *Suspicion* balances the dread of Lina's unraveling mind with the absurd humor that Hitchcock is known for. It probably helped that Alma Reville, Hitchcock's partner in life and work, cowrote the script with Joan Harrison and Samson Raphaelson. Harrison also had a long working relationship with Hitchcock, as she coproduced *Alfred Hitchcock Presents* (1955–1962) and had coadapted *Rebecca* and *Jamaica Inn*.

As a cautionary tale, *Suspicion* portrays Lina as not wrong in every appraisal of her husband, whom she married perhaps a bit too soon. Johnnie, it turns out, is a broke gambler who lies, steals, and has little regard for Lina's feelings. She might be wrong about him being a murderer, but as a modern audience, we certainly can't blame her for thinking so, as Johnnie started their relationship with forceful wrist pulling and nonconsensual touching. Ick. While this sort of thing wasn't as odd in the early forties, it doesn't age well.

All that said, we both loved watching the rollercoaster ride of this marriage depicted by talented actors and are only disappointed by the ending, which proves Lina wrong about the murdering. In fact, Johnnie is suicidal over his stealing money from his boss (and cousin), and the film ends with them vowing to put their marriage back together. Because the movie is perfectly wrought with suspicion, as the title denotes, the ending feels mismatched, especially if you know the plot of the novel it's based on.

> In 2023, films based on books made up 70 percent of the top-twenty-grossing films worldwide and generated 53 percent more revenue than those made from original screenplays.[1]

Before the Fact (1932) by Anthony Berkeley Cox, under the pseudonym Frank Iles, is a psychological suspense British novel. *Suspicion* is based rather loosely on *Before the Fact*, as in the film Johnnie is innocent of murder, and in the novel that is less obvious, though not outright confirmed. So, why did the master of suspense take out the murder and end his film with an attempt at a happy ending? It was because both the studio, RKO Pictures, and those working for Cary Grant weren't happy about his portrayal of a murderer on screen. RKO had come up in the Golden Age of Hollywood, when stars' reputations were vital to the success of the pictures he or she starred in. Casting Cary Grant in *Suspicion* made for a dynamic character but pushed Hitchcock into what he called "compromises." In a BBC broadcast that aired in 1969, Alfred Hitchcock spoke about the making of *Suspicion* under RKO's reign:

> In the case of *Suspicion*, one of the early films you mention with Cary Grant, he should never have been in the picture in the first place. You run into this problem. You cast a man who is suspected

of murder and then you have to compromise. I remember the head of RKO returned from New York and said, with a big grin on his face, "Oh, you should see what's been done to your film *Suspicion*." I said, "What?" He said, "Wait and see." It was now only fifty-five minutes long. They had gone through the film in my absence and taken out every scene that indicated the possibility that Cary Grant was a murderer. So, there was no film existing at all. That was ridiculous. Nevertheless, I had to compromise on the end.[2]

This push and pull between a director, especially one who works as an auteur, and a studio happens all the time. Many film endings have been changed due to focus groups, in which audiences give feedback for a film not yet released, or by the studio itself. An example is one of our favorite horror comedies, *Little Shop of Horrors* (1986). In the original cut, both Seymour (Rick Moranis) and Audrey (Ellen Greene) are eaten by the hungry human-eating plant, Audrey II. But, according to director Frank Oz, test audiences hated it. So, it was back to the drawing board to create a happy ending in which the couple kill Audrey II and live happily ever after. We agree—it's a more satisfying ending.

In his interview with François Truffaut, Hitchcock outlined how he would've liked to end the film, if RKO, and Grant's representation, hadn't been so averse to Grant's villainy on screen:

I'm not too pleased with the way *Suspicion* ends. I had something else in mind. The scene I wanted, but it was never shot, was for Cary Grant to bring her a glass of milk that's been poisoned, and Joan Fontaine has just finished a letter to her mother: "Dear mother, I'm desperately in love with him, but I don't wanna live because he's a killer. Though I'd rather die, I think society should be protected from him." Then Cary Grant comes in with the fatal glass, and she says, "Will you mail this letter to mother for me, dear?" She drinks the milk and dies. Fade out and fade in on one short shot: Cary Grant whistling cheerfully, walks over to the mailbox and pops the letter in.[3]

And that's why he's the most revered suspense director of all time. Even though some studios and test audiences make for better endings, we do feel that *Suspicion* could've been even better with the ending Hitchcock outlined. It has that touch of irony (Johnnie mailing the letter not knowing it will be his demise) that feels uniquely Hitchcockian.

The character of Isobel Sedbusk, played by Auriol Lee, is another aspect of *Suspicion* that feels particular to Hitchcock's style. She is a murder-mystery writer, à la Agatha Christie, who knows *a lot* about poison and murder in general. She is quite gleeful on the subject, not knowing how troubled her friend Lina is about it. Isobel owns numerous books on poisons; in fact, she lent one to Johnnie, which makes Lina suspect her husband even more. This character with a wealth of interest and knowledge in murder is seen again, twofold, in our next-covered Hitchcock film, *Shadow of a Doubt* (1943). Joseph Newton (Henry Travers) and his neighbor Herbie Hawkins (Hume Cronyn) talk endlessly about true crime, as well as read crime fiction. They would undoubtedly enjoy a chat about untraceable poisons with the mystery author in *Suspicion*, Isobel. Now, that's a crossover we'd love to see!

Unfortunately, actress Auriol Lee, who portrayed Isobel, was driving cross-country from New York City to Hollywood after she filmed *Suspicion* and was killed in a car accident in Kansas at the age of sixty. A legend of Broadway and the West End, Lee had only been in two films: *A Royal Divorce* (1938), in which she plays Napoleon's mother, and *Suspicion*, which she never got to see.

In *Suspicion*, Alfred Hitchcock's cameo is another blink-and-you-miss-it appearance. We hadn't caught him with our own eyes, so we had to consult The Hitchcock Zone to discover he is mailing a letter in a mailbox next to Mrs. Newsham (Isabel Jeans). Maybe it's a nod to how he originally wanted to finish the film?

Johnnie (Cary Grant) is bothered by the scent of a cigar in the next compartment in the film. What is the biology behind sensitivity to smell? Although many of us can be bothered by everyday scents, those with hyperosmia have a heightened sense of smell, and perfumes, scented candles, and cleaning products could cause illness or migraines. Pregnant women often experience hyperosmia, especially in the first trimester, and this can lead to worse morning sickness. People with autoimmune diseases and other conditions may also have a heightened sense of smell.[4] The best way to treat hyperosmia is to avoid strong scents altogether. That is easier said than done when someone is oblivious to the smell of his cologne! Making others aware of the condition will hopefully lead them to be sensitive toward those who may react to smells.

Johnnie is told to smile for the photograph being taken in *Suspicion*. How is happiness shown in various cultures across the world? In the United States it's typical to smile, with or without showing teeth, and others will understand the emotion. In Russia, smiles are saved for close friends and family, while in Japan, it may be considered rude to show your teeth when smiling. Before traveling, it's worth researching so you can either fit in to the culture you're visiting or at least understand why someone may give you a rude look if you're grinning too broadly.

Johnnie is trying to flirt with Lina, but his techniques are awkward. What is the psychology behind flirting, and how does it compare with animals finding mates? Humans flirt using verbal and nonverbal communication, and it is instinctual, not logical. Research suggests we flirt with those who we believe are like us in some way, and the act itself increases dopamine and boosts self-esteem.[5] Flirting can be as simple as prolonged eye contact or can be complicated like taking part in intimate conversations. In the animal world, there are complicated ways to woo a potential mate. "Male puffer fish work nonstop for a week to construct incredible 'crop circle' art to attract the attention of passing females,"[6] while female frogs will choose a less appealing call if presented with a strong call, a weaker one, and a decoy.[7] What does this teach us? Love is complicated! And flirting can be awkward.

Speaking of flirting, how has courting changed throughout history? Dating didn't really exist in the distant past because marriages were often arranged and treated more like a transaction—not very romantic. It took centuries for women to gain equality and to be seen as individuals. Depending on culture, class, religion, and expectations, dating can look vastly different all across the world. We discovered that "scientific research into courtship began in the 1980s, after which time academic researchers started to generate theories about modern courtship practices and norms. Researchers have found that, contrary to popular beliefs, courtship is normally triggered and controlled by women, driven mainly by non-verbal behaviors, to which men respond."[8] Let's go, girls.

Lina and Johnnie elope in *Suspicion*. In a 2023 poll, 62 percent of couples surveyed were considering elopement. Reasons included wedding cost and a desire for a more intimate ceremony.[9]

The waltz they keep dancing to throughout *Suspicion* brings up the memory of the first time they met. How does memory work specifically with music? We all can recall a feeling that a particular song brings us: a childhood memory, our first love, a breakup, or witnessing a band perform the song live. Music evokes both positive and negative occurrences in our lives and is a powerful tool being used by professionals:

> Researchers at the music and neuro-imaging laboratory at Harvard-affiliated Beth Israel Deaconess Medical Center have shown that singing lyrics can be especially helpful to people who are recovering from a stroke or brain injury that has damaged the left-brain region responsible for speech. Because singing ability originates in the undamaged right side of the brain, people can learn to speak their thoughts by singing them first and gradually dropping the melody. Former Representative Gabrielle Giffords used this technique to learn to speak well enough to testify before

a Congressional committee two years after a gunshot wound to her brain destroyed her ability to speak. Singing has also helped healthy people learn words and phrases faster.[10]

Music has also been shown to be helpful for Alzheimer disease and dementia patients by calming them, improving memory recall, and encouraging a positive atmosphere. Patti Naiser, who helps seniors transition from their homes, said, "Music is an expression of humanity and is, in and of itself, healing. It's something we all understand and something we are all touched by."[11]

Johnnie has a lot of red flags, so to speak, in his relationship with Lina. What do relationship experts view as signs of potential problems? We all have turning points during dating that can make or break a relationship. For example, if you meet your partner's friends for the first time and the partner acts completely out of character and is cruel to you, this could be a red flag. Other potential problems include jealousy, incompatible goals, verbal or physical abuse, a history of cheating, excessive substance abuse, and being controlling.[12] If you notice any of these issues, communicate your concerns with your partner, and don't stay in an unhealthy relationship.

One red flag Johnnie has in *Suspicion* is a gambling addiction. This is considered a disorder by medical professionals and shares traits similar to drug or alcohol dependence. People who are gambling addicts are constantly thinking about it, planning their next outing, increasing the amounts they bet, getting a high from the act, and lying about gambling. This can cause relationship problems at home and work and can lead to significant financial issues. This addiction can be treated with therapy, self-help groups, and medications.[13]

Beaky (Nigel Bruce) trusts Johnnie, probably because we are all inherently born with a truth bias. We believe people have our best interests at heart and are telling us the truth. It's not until we are lied to often that our truth bias turns into a lie bias. Those incarcerated in prisons tend to have a lie bias. They don't trust that others are telling them the truth and are more wary of truths.[14] There is also a tendency in communication to experience the halo effect. This means if you perceive something positive about a place or person, you'll assume all things about such are good. The reverse halo effect exists when you perceive something negative and postulate that everything about the person, place, or product is bad. A way to combat these tendencies is to get to know people beyond a first impression, visit a place more than once, and give products or other things another try to assess them more accurately.

Johnnie speaks to a mystery writer about untraceable poisons, which reminds us of the queen of mystery, whom we wrote about in our 2023 book, *The Science of Agatha Christie*. The author mentions Richard Palmer, who murdered someone with brandy, and discloses that she loaned Johnnie the book two weeks prior. Lina passes out when they get home and sleeps for hours. She learns that the author revealed to Johnnie a poison that is common and untraceable after death.

This gives Lina pause to consume a glass of milk she is given before bed by her husband. There are many poisons that people have used to murder others in the past, including a 2023 case of a man using a lethal dose of colchicine to kill his wife. The drug is used to treat gout, and the victim, Betty Bowman, did not have the condition and had not been prescribed the medication. Her husband, a former doctor, tried to order an immediate cremation after her death, but the medical examiner's office halted the order, citing suspicious circumstances. Computer records showed Connor Bowman had researched the drug and converted his wife's weight to kilograms to ascertain a lethal dose of the drug for her size.[15] As of this writing, Bowman is awaiting trial for first- and second-degree murder charges.

Studio executives made the decision that viewers wouldn't have liked the ending Hitchcock, and we, would have preferred. Thankfully, in the modern era, we can see our favorite actors portray all sorts of roles, and it doesn't necessarily change our opinions of them.

CHAPTER SEVEN
Shadow of a Doubt

Though the screenplays he worked from were vital to telling suspenseful stories, Alfred Hitchcock's direction is what truly makes his films stand out, even over half a century later. We'll start with a quote about *Shadow of a Doubt* from one of the most influential film critics of the twentieth century, Roger Ebert:

> Much of the film's effect comes from its visuals. Hitchcock was a master of the classical Hollywood compositional style. It is possible to recognize one of his films after a minute or so entirely because of the camera placement. He used well-known camera language just a little more elegantly. See here how he zooms slowly into faces to show dawning recognition or fear. Watch him use tilt shots to show us things that are not as they should be. He uses contrasting lighted and shadowed areas within the frame to make moral statements, sometimes in anticipation before they are indicated. I found while teaching several of his films with the shot-by-shot stop-action technique, that not a single shot violates compositional theory. Not many directors were fonder of staircases than Sir Alfred. They impose a hierarchy of power and weakness. A character at the top of the stairs can seem to loom or be in danger of toppling, depending on whether the POV is high or low. The flow at the house goes up the sidewalk, onto the porch, through the door and directly up the stairs. There are outside stairs in the back, and both staircases are used for tight little sequences of threat and escape. Notice how many variations of camera angles and lighting Hitchcock uses with the stairs. He considered them an ideal device for introducing imbalance into otherwise horizontal interiors.[1]

Translating the language of film to text is not easy, but Roger Ebert does a masterful job here in conveying what makes Hitchcock's eye so unique.

Unlike many of Hitchcock's films in which an innocent man is accused of murder (like the aforementioned *Suspicion* and *The 39 Steps*), Charlie Oakley (Joseph Cotten) is, indeed, guilty. This is obvious from the outset, as his dark mood and mysterious manner follow him as he finds refuge with his sister, Emma Newton (Patricia Collinge), as he is being pursued by lawmen. Also, just something in the casting of Joseph Cotten makes for a killer. Unlike Cary Grant, who was trying to save his reputation as Hollywood's debonair good guy, Cotten embraced darker roles. Meg remembers him vividly from one of her favorite horror films from childhood, *Hush . . . Hush, Sweet Charlotte* (1964), in which he plays a deviously charming doctor with murder on his mind. Because he was a good friend to Orson Welles, Cotten appeared in career-making turns in *Citizen Kane* (1941) and *The Magnificent Ambersons* (1942). Cotten also starred in the popular *Gaslight* (1944), which we discuss in our book *The Science of Women in Horror* (2020) as the genesis of the term *gaslighting* due to the psychological abuse depicted in the film.

Charlie (Teresa Wright), named after her beloved uncle, Charles, is the film's young protagonist. Although she is young and innocent, "the slight-looking Young Charlie also gets to display unbending inner strength, even risking her life to confront her psychopathic blood-relative."[2] She is a formidable precursor, clever and quick on her feet, to the more sophisticated heroines that will arrive in Hitchcock's films in the characters portrayed by the likes of Grace Kelly and Tippi Hedren.

"In choosing a name, research finds that parents strive both to be unique and 'fit in' with the crowd. They try to choose names that will help their child to stand out yet also will seem somewhat familiar. Accordingly, they often choose a slightly different version of a popular name from previous years."[3]

One of our favorite parts of *Shadow of a Doubt* is Charlie's little sister, Ann, played by Edna May Wonacott. Ann provides much needed levity as a precocious bookworm who is too busy reading to care about all the drama going on. In a stroke of luck, nine-year-old Wonacott walked up to Alfred Hitchcock and producer Jack Skirball at an intersection in Santa Rosa, California. They were immediately taken with her and asked if she would like to be in a movie filming in town. She said yes, and later that day Hitchcock and Skirball showed up on her parents' doorstep asking for their daughter to do a screen test in Hollywood! We're still waiting to be discovered on a street corner by a world-renowned director.

Wonacott signed a seven-year contract with the studio, yet her acting career was short-lived. Her last credit is a 1952 episode of the spy adventure series *Biff Baker, U.S.A.* (1952–1954). Wonacott chose to raise a family instead of pursuing stardom. In a 2012 interview, she spoke of her time working with Hitchcock, as well as befriending his daughter, Patricia. She also ruminated on why *Shadow of a Doubt* still resonates with audiences:

> It amazes me that the film is still so popular, and everybody seems to remember it. It is really one of the true classics and I feel really proud to have been a part of it. I carry really fond memories for that part of my life, and I am thankful to have such a blessed time of my life on film, so to speak. I also have lots of scrap books and have been invited to lots of functions to show my scrap books and share that part of my life. In fact, I still receive fan mail and a lot of it in the last few months has been coming from Europe. Lots of people in their thirties and forties are turning to the classic films nowadays.[4]

Known as the "Merry Widow Murderer," Charlie Oakley has a penchant for killing rich widows whom, we come to find out later in the film, he holds a violent resentment toward. Naturally, we wanted to find out if there was a real-world inspiration for Wilder's fictional killer and were pointed toward Earle Nelson. Nelson had several monikers, including

"Gorilla Man, "Gorilla Killer," and "Dark Strangler." Like the fictional Merry Widow Murderer, these nicknames have a silly sort of ring to them and were popular for newspaper headlines. But don't let them fool you; Earle Nelson was a human monster. From an early age it was clear that something was off with Nelson, who was born at the cusp of the twentieth century. He was orphaned at a young age and spent time in a mental hospital before his murderous rampage that ended with his hanging in Winnipeg, Canada, in 1928. His crimes were outlined on the popular blog *Murderpedia*:

> Starting in San Francisco and working his way up the coast to Seattle, he headed eastward. In his wake, the papers dubbed him the "Gorilla Man"—a nickname that had less to do with his appearance (tho' he was no Clark Gable) than with the animal savagery of his attacks. His targets were mainly middle-aged or elderly landladies who had placed "Rooms to Let" ads in the paper. Nelson (who could be a charmer when he needed to be) would show up and ask to see a room. Once alone with his prey, however, he would undergo a classic Jekyll/Hyde transformation.
>
> Typically, he would choke the women to death, commit post-mortem rape, then stash the corpses in strange hiding places. One of his victims was stuffed unceremoniously into an attic trunk, while others were crammed into basements and behind furnaces. The final victim was discovered by her husband *as he knelt for his nightly prayers*. His bride was crammed beneath their bed.[5]

Nelson is known to have killed at least twenty-one women, the vast majority of them being aged in their forties to sixties. He continued his killing into Canada, where he was eventually caught. Although Wilder and his fellow screenwriters certainly couldn't emulate all of Nelson's habits (particularly the sexual nature of his crimes), it seems that his penchant for killing older, widowed women was an inspiration. This dichotomy of the brutal truth and the more palatable Hollywood version is resonant of Hitchcock's later film *Psycho* (1960), based on the killings (and grave robbing) done by Ed Gein.

Fascinatingly, *Shadow of a Doubt* is the only film written by the immensely successful playwright Thornton Wilder. A Pulitzer Prize winner both for his novel *The Bridge of San Luis Rey* (1927) and for his play *Our Town* (1938), Wilder was a formidable writer who used his talents to craft a screenplay with both humor and suspense (Hitchcock's bread and butter). He was assisted by Alma Reville, along with fellow first-time screenwriter Sally Benson, who went on to write such notable films as Elvis's *Viva Las Vegas* (1964) and the biopic *The Singing Nun* (1966).

> Where is Alfred Hitchcock this time? Only about fifteen minutes into the film, you can spot the director playing cards on the train Charles Newton is riding to escape the law. Because the camera is angled over his shoulders, we can see he has a straight flush of spades in his hands!

Mr. Spencer is told he looks tired and is encouraged to take a nap. He doesn't, but what is the science behind napping? Studies have shown that shorter power naps can boost energy and alertness, while naps longer than an hour are linked to obesity and increased cardiovascular diseases.[6] Say it isn't so! We both love taking long naps. Thankfully, more research has been done that gives more encouraging results for those of us who enjoy a midday catnap:

To better understand connections between daytime naps and health, researchers assessed data taken from more than 3,200 adults living in Spain, a country where midday naps, or siestas, are common. They found that about one-third of adults took regular siestas—oftentimes four days a week. Among regular nappers, those who snoozed for thirty minutes or less were 21% less likely to have elevated blood pressure compared to non-nappers. Those who napped for more than thirty minutes were more likely to have a higher body weight. They were also 41% more likely to have high blood pressure, high blood sugar, and a larger waist circumference. Upon further analysis of the data, the researchers found that certain activities, including going to bed later, smoking, and having larger lunches later in the day, helped explain links between longer naps and increased cardiovascular disease risks.[7]

We volunteer for any further testing needed.

Mr. Spencer sends a telegram in *Shadow of a Doubt* that seems vaguely like a code. What is the science of codes? "The preparation and analysis of ciphers and codes—cryptography and cryptanalysis—together make up the science of cryptology (from the Greek *kryptos*, meaning 'hidden,' and *logos*, meaning 'speech')."[8] The first cipher was used in 400 BCE by the Greeks for secret messages between military commanders, and technology has evolved to encrypt data through microprocessors to protect information being sent digitally.

Charlie is in a funk in the beginning of the movie, feeling like life is mundane and redundant. How can we get ourselves out of a rut when we feel like this? Doctors recommend reaching out to a friend, engaging in some self-care, and eating some dark chocolate. Chocolate made of 70 percent cocoa or more contains flavanols that help you think more clearly by increasing blood flow to your brain.[9] Sounds good to us! Listening to music, going for a walk, and reading a good book can all help you feel better about life and yourself.

The kids in the house are all talking at once in the movie, and it's quite comical. How much do kids talk in the modern era? Everyone is different, of course, but a recent poll found that the average child asks eleven questions a day, which makes that a whopping four thousand questions per year! Now, multiply that by the number of kids in your house and you're looking at a lot of talking. It's not all bad, though. Parents report that they learn a lot from their children's questions—about their curiosity and knowledge, and parents gain insight into new topics.[10]

It's discussed in *Shadow of a Doubt* that the youngest child in a family is usually the most spoiled. Is this true? Alfred Adler, a psychologist who studied birth order, found that the youngest sibling tends to have self-confidence, be funny, and be creative but can also misbehave, active impulsively, and be less independent.[11] The study also found that youngest children spend more time with the parents on average, so it can be perceived that they are spoiled, but it's not always the case. Every family is different!

Telepathy is mentioned in this film. Is there science to prove it? Not yet. There have been numerous studies, experiments, and tests to try to show that telepathy exists and many personal anecdotes and stories seem to suggest it's possible, but some experts disagree. Experiments haven't always had the proper controls and lacked repeatability. As Felix Planer wrote in his book *Superstition*, "Many experiments have attempted to bring scientific methods to bear on the investigation of the subject. Their results, based on literally millions of tests, have made it abundantly clear that there exists no such phenomenon as telepathy, and that the seemingly successful scores have relied either on illusion, or on deception."[12] Even if telepathy hasn't been proven to be true by science, we still appreciate it as a plot point in books, television, and movies.

Joe tells Charles not to put his hat on the bed. What is the superstition behind this? There are many superstitions regarding hats, and this one comes from the belief that evil spirits live in our hair. Most likely this came from the fact that static-electricity buildup causes hair to "float" when taking off hats in a dry environment. The hats, then, must contain evil and should not be left where we sleep. Another theory to this belief is that a hat left on a bed may signify that someone has died. This could inadvertently conjure spirits![13] We're going to play it safe and not put our hats on beds from now on.

Charles proclaims that he can't speak before having his coffee in the morning. How many people feel like this? Meg certainly does! Caffeine is the most widely used mood-altering drug in the world, with an estimated 80 percent to 90 percent of children and adults consuming it regularly.[14] In this study, 26 percent of people reported that they needed caffeine to function. How does caffeine affect us? Since it is a mild stimulant, it can increase our feelings of happiness, energy, and well-being but could cause health problems like hypertension, anxiety, insomnia, and cardiac issues.[15] We're going to stick to our coffee but take this knowledge under advisement.

Mrs. Newton is adamant about the order of ingredients as she is baking a cake. She is probably aware that science is behind baking. The definition of a *chemical reaction* is "a process in which one or more substances are converted to one or more different substances." As we mix ingredients for a cake, for example, the dry items dissolve into the wet, and this creates a batter. In chemistry, this is called an aqueous solution. The reactions of ingredients during the baking process cause the cake to rise and expand while the cake solidifies due to the gluten strands drying out. A perfect, golden-brown crust is created as the result of a chemical reaction called the Maillard reaction.[16] Science is everywhere, and now we're hungry.

In *Shadow of a Doubt*, Ann reads two books a week. Meg averages this number as well, while Kelly's reading mainly consists of grading 150- plus papers a week. What are the statistics regarding how much the average person reads? According to a report from 2023, the average person reads a little over twelve books a year, with print books being favored over audiobooks.[17] Why should we read? We realize we're preaching to the choir if you're reading this, but know that it exercises the brain, improves concentration, increases your knowledge, reduces stress, and increases your empathy.[18]

Herb brings mushrooms to the house in the film, and Joe talks about killing someone in a bathtub. The mother claims it's Joe's way of relaxing. Why are some people comforted by true crime and serial killers? Recent studies have found that women, specifically, are more likely to watch, read, and listen to content focusing on true crime. Professor David Schmid said, "This is a trend that goes back a long way. Much earlier in the 20th century when people started to realize a majority of crime fiction readers are women, it was met with a similar kind of surprise and shock. But it makes sense that women would be more captivated by this content as they are more often than men the victims of violent, intimate crimes. The interest in the subject comes from women's awareness that they are the victims of a vast majority of interpersonal violent crime especially crimes like serial murder."[19]

"Each year nearly a thousand people are killed in train related accidents. More than half of all railroad accidents occur at unprotected crossings. More than 80% of crossings lack adequate warning devices such as lights and gates."[20]

Uncle Charlie ends up being a douchebag and dies under a train. Good riddance! He did try to murder his loved ones with car exhaust and in a garage. How long would it take for someone to perish like that? Carbon monoxide poisoning is the leading cause of fatal poisonings in North America and can kill in only a matter of minutes. Make sure not to run vehicles in an enclosed space, open doors or windows if there is a carbon monoxide leak, and get a detector to keep you safe.[21]

CHAPTER EIGHT

Spellbound

It was Christmas morning, and Santa had left me (Meg) a special treat under the tree. It was a four pack of early Alfred Hitchcock films on VHS—something every twelve-year-old girl would want, right? Okay, maybe it was not *every* sixth grader's ideal present, but I was thrilled to add to my growing Hitchcock collection. This was the first time I got to watch *Spellbound*, which to this day is tough to find on streaming. Thinking back on unwrapping that gift and discovering films like *Spellbound* and *Rebecca,* I find writing this book to be particularly poignant.

Like many films directed by Alfred Hitchcock, *Spellbound* is based on source material. In this case, it's the 1927 British novel *The House of Dr. Edwardes.* Written by Hilary A. Saunders and John Palmer under the shared pseudonym Francis Beeding, *The House of Dr. Edwardes* reads similarly to the plot points of *Spellbound.* The first chapter begins with a description of the main character: "Constance Sedgwick, M.D., aged twenty-six, was staring at herself critically in the long mirror. As a young Doctor of Medicine, with a degree for which she had worked hard and long, she prided herself on being objective. She was looking at herself, so she said, as she had been taught to look at a bacteriological culture under the lens, very steadily and without prejudice."[1]

Spellbound begins with Constance (her surname switched to Peterson), portrayed by Ingrid Bergman, looking similarly scientific. Wearing glasses and a white coat, she is not the typical 1940s-era bombshell with a ready smile. In fact, when she is sexually harassed by her boss (*ick!*), Dr. Peterson reacts with a frown and little mirth, unlike many of her contemporaries who may have felt uncomfortable in the era. *Spellbound* has moments of fledgling feminism that we admire, like when Dr. Peterson is sitting on a couch in the Empire State Hotel ocre.

and is being harangued by a man (Wallace Ford), only to be "saved" by the house detective (Bill Goodwin) and then harassed by him too. It's an irony that resonates, especially as Dr. Peterson has worked tirelessly to be a respected woman of psychiatric medicine.

Ingrid Bergman is considered one of the best actresses in cinema, enhanced by her impressive stack of Academy Awards wins and nominations. These include her two wins in the Best Actress category for 1945's *Gaslight* (costarring *Shadow of a Doubt*'s Joseph Cotten) and as a woman posing to be a deceased grand duchess in *Anastasia* (1956). Bergman made three films with Hitchcock: *Spellbound*, *Notorious* (1946), and *Under Capricorn* (1949). While there was known to be rifts between Hitchcock and some of his other ingenues, he and Bergman became lifelong friends until Hitchcock's death in 1980, before which Bergman was said to give Hitchcock comfort in his final days. She died only two years later.

One instance from the set of *Spellbound* shows the relationship between Hitchcock and Bergman:

> When she ran into some trouble performing a scene, Bergman once got into a brief argument with Hitchcock. She explained to the director that she "didn't think [she] could do it naturally," and he simply responded, "All right. Fake it." This advice of course bears a stark contrast to the 'Method' approach, which encourages actors to draw upon real-life experiences to produce emotions for a given scene. *Spellbound* leading man Gregory Peck pointed out the incompatibility this would pose with Method actors like Marlon Brando. "He never asks you to dredge up old emotions, he wants an effect," Peck explained, adding that "Method actors might find his technique unsettling."[2]

The method that Peck refers to is an acting process developed by acting coach Lee Strasberg for his American students, with inspiration from his training by Russian legend of acting Konstantin Stanislavsky. At the website for The Lee Strasberg Theatre & Film Institute you can find a description of the method:

At its core, Method Acting is therefore a systematic approach to training the living material that is the actor's 'instrument,' as well as a means for preparing a role. The use of Lee Strasberg's exercises both develop the content of the actor's talent and provide a roadmap to the individual's creation of a character. The use of one's own life experiences in the creative imagination infuses each choice with genuine thought, desire, sensation, action and feeling resulting in psychologically in-depth behavior. It builds upon the work of Stanislavsky, and as Lee believed, accomplished what Stanislavsky set out to achieve.[3]

Gregory Peck mentioned Marlon Brando as a method devotee, while many others learned and utilized Strasberg's method of connecting their own experience to their work, including famous actors such as Dustin Hoffman, Robert De Niro, and Marilyn Monroe. Although Strasberg died in 1982, his method lives on, with notable actors like Daniel Day-Lewis and Jared Leto using his teachings. Currently, branches of Lee Strasberg's institute are in Los Angeles and New York, in addition to international workshops in places such as Tel Aviv and Copenhagen. You can even learn the method online. The actors that we interviewed for our 2021 book, *The Science of Serial Killers*, don't recommend using this process, as it may lead to problems with others and within yourself.

Hitchcock's advice for Ingrid Bergman to fake it would certainly ruffle a method actor's feathers, as Peck stated. This may be one reason Hitchcock was known to be a difficult director to work with—difficult but effective.

This was the only film collaboration of Gregory Peck and Hitchcock. Peck is best known for his Oscar-winning turn as Atticus Finch in *To Kill a Mockingbird* (1962) and for starring opposite Audrey Hepburn in the romance *Roman Holiday* (1953). In *Spellbound* he plays John Ballantyne, a man with a bad case of amnesia (and possibly a murderer) who not only needs Dr. Peterson's expertise in psychoanalysis but also falls madly in love with her. Dr. Peterson feels the same, and we can't say we blame her. Despite his character's flagrant red flags (maybe even worse than Cary Grant's in *Suspicion*), Gregory Peck is so damn handsome.

> Science is behind whom we find attractive! For the
> face specifically, symmetry, lack of blemishes, and
> expressiveness influence others.[4]

Deeply rooted in the work of Sigmund Freud, Dr. Peterson speaks of the popular psychiatry of the 1940s, like the Oedipal complex, because, you know, we're all attracted to our parents! And when Dr. Peterson and John share their first kiss there is a symbolic shot of many doors opening, as if this kiss has opened a part of their brains that was not there before. Because the film deals with the complex inner workings of the mind, Hitchcock was inspired to depict these with the help of a famous artist. This was most likely the final straw in the relationship of producer David O. Selznick and Hitchcock. They were obligated to work one more time after *Spellbound,* on *The Paradine Case* (1948), which is one of Hitchcock's lesser-known and lesser-regarded films. Here's the story:

> Hitchcock had the brilliant idea of using surrealist legend Salvador Dalí to design the sequences of delusion. Selznick was initially unsure, but when he realized the publicity boon it would cause, he agreed. However, things quickly went downhill: Hitchcock gave Dalí pretty much free rein, and had little to do with the sequence, and as a result, Selznick tried to bring it under control by hiring legendary production designer William Cameron Menzies (*Gone with the Wind*) to helm the sequence — although Menzies was unhappy with the end result, and had his credit taken off. Once it was completed, it seemingly ran as much as twenty minutes (according to Ingrid Bergman in Donald Spoto's book *The Dark Side of Genius: The Life of Alfred Hitchcock,* 1999), but Selznick cut much of what remained (including shots of Bergman as the god Diana). Ultimately, fourteen minutes of the film in total were removed by the producer (which seems to have been lost).[5]

We have to wonder what those Dalí sequences were like in their full glory. Too bad they are lost to history. Spanish artist Salvador Dalí was known as a surrealist whose work would convey surreal visuals, like his 1931 painting *The Persistence of Memory*, which we can nearly guarantee you've seen. It's the one with the melting clocks on the beach. You can see the original at the Museum of Modern Art in New York City. If you don't know his art, then you might know Dalí, himself. He was known for his eccentrically thin moustache and has been portrayed several times on screen. Adrien Brody played him in *Midnight in Paris* (2011), and Dalí's biopic *Daliland* (2022) stars Sir Ben Kingsley.

We're happy to report that we caught Alfred Hitchcock in his cameo in *Spellbound*. He is first in line coming out of an elevator of the Empire State Hotel, where Dr. Peterson is searching for John. Hitchcock puffs on a cigar as he quickly exits, leaving the acting to the actors.

What is dissociative amnesia as portrayed in the film? According to the journal *Psychiatry*, "Dissociative amnesia is a disorder characterized by retrospectively reported memory gaps. These gaps involve an inability to recall personal information, usually of a traumatic or stressful nature. Dissociative amnesia most commonly occurs in the presence of other psychiatric conditions, particularly personality disorders."[6] People may experience memory loss through other means like traumatic brain injuries, lesions, or toxic exposure. Dissociative amnesia is characterized by the inability to recall personal information of a stressful nature.

Nervous exhaustion is also mentioned in *Spellbound*. How is it treated? If you are mentally exhausted, it means you've been focusing on a tough task for too long. You can get irritated, have trouble focusing, feel overwhelmed, have a hard time sleeping, and cope by using drugs or alcohol. All of this can lead to depression and anxiety.[7] It's important to take breaks from difficult tasks, relax when you can, and step away to get some alone time. Taking a mental-health day is encouraged so you don't burn out.

Dr. Edwards seems to have a phobia of parallel lines when Constance traces some on the dinner table. What causes this, and what are some other phobias? We discover by the end of the film that the parallel lines brought up a lost memory of the ski slopes. Other phobias aren't always linked to an event. The most common phobias are arachnophobia (a fear of spiders), claustrophobia (a fear of enclosed spaces), and trypanophobia (a fear of injections). Less common ones include sesquipedalophobia (ironically the fear of long words), omphalophobia (the fear of belly buttons), and chaetophobia (the fear of hair).[8]

> According to Johns Hopkins Medicine, approximately nineteen million Americans have one or more phobias that range from mild to severe.[9]

A patient in *Spellbound* is suffering from a guilt complex, believing he killed his father. Later, Dr. Edwards is diagnosed with the same. What do medical professionals say about guilt complexes? Although it's natural to feel guilty about something you regret, it can become a problem when you fixate on the emotion for a long time. Long-term guilt can lead to depression, anxiety, and low self-esteem and harm relationships. The best way to overcome a guilt complex is to move forward, either through therapy or by forgiving yourself.[10] We are unable to change the past, but we can assure that we don't make the same poor decisions in the future.

Could our dreams be premonitions, as implied in *Spellbound*? While the scientific jury is still out on whether our dreams can predict the future, there are quite a few anecdotes throughout history that suggest otherwise. Abraham Lincoln purportedly dreamed of his own assassination. Sleep Foundation.org reports, "Lincoln dreamed of people sobbing, and in his dream, he went to investigate. In the East Room of the White House, he found a corpse dressed for a funeral. Lincoln asked the figures in the dream what happened. One reported to him that the president was assassinated."[11] He told his friend and law partner about the dream just days before his death. In 1966 seventy-six separate people reported premonitions of a landslide that occurred in the South Wales village of Aberfan. Waste from a coal mine slid into the village and destroyed the school, killing 144 students and teachers. One girl, before the accident, dreamed her school was gone and had been covered by something black.[12] While scientists explain these dreams by selective memory recall or coincidence, others believe in the supernatural.

The truth (and love) prevails at the end of *Spellbound* and the audience is left to believe the couple will live happily ever after. We were on the edge of our seats, but we should have known that Hitch wouldn't let us down!

CHAPTER NINE
Strangers on a Train

As a film-studies minor in college, I (Meg) was introduced to many films and directors who changed the landscape of cinema. I watched Akira Kurosawa's *Rashomon* (1950), Ingmar Bergman's *The Seventh Seal* (1957), and even super racist but important to film history *The Birth of a Nation* (1915), directed by D. W. Griffith. To study Alfred Hitchcock's contribution to movies, my professor chose 1951's *Strangers on a Train*. The film was profiled in Cinephilia & Beyond:

> This captivating story of two people meeting on a train and conversing about the execution of a perfect murder has forever remained a much-desired topic of analysis and debate among film enthusiasts all over the world. What distinguishes *Strangers on a Train* from similar films, even within Hitchcock's own canon, is the fascinating idea at the center of it—the motif of doubles, the inner battle of good and evil in all human beings—as well as impressive technical virtuosity we grew accustomed to when talking about the works of [Hitchcock]. The suspense is so powerful it can be felt though the screen.[1]

It becomes obvious that Bruno is a psychopath when he takes it upon himself to kill the estranged wife, Miriam (Kasey Rogers), of the other conversant (the character Guy Haines). Hitchcock conveys Bruno's malicious intent with the use of shadow as he stalks Miriam through an amusement park. When he catches up to her, Bruno strangles Miriam, as we watch through a reflection of Miriam's fallen glasses. These glasses end up being a pivotal plot point, an artful way to show the murder and symbolize the before-mentioned theme of doubling, or mirroring.

71

Strangers on a Train is based on the debut novel of the same name by Patricia Highsmith. Several of Highsmith's works would later be adapted into film, including *The Talented Mr. Ripley* (1999), starring Jude Law and Matt Damon, and the more recent queer romance *Carol* (2015), starring Cate Blanchett and Rooney Mara. Hitchcock was a fan of Highsmith's novel and chose *Strangers on a Train* for his project, rather than being assigned by the studio.

Nonetheless, the writing of the script proved to be a challenge, as novelist Raymond Chandler was hired for the task and did not gel well with the auteur director. Yes, this was yet another feud in Hitchcock's quest for cinematic perfection. Chandler sent two drafts of a screenplay to Hitchcock, who never responded back to the author, making for an awkward relationship. Also, Hitchcock essentially eviscerated Chandler's scripts, making them his own. This did not sit well with Chandler, who was a well-known writer with an ego. He wrote to the director after seeing *Strangers on a Train*, lambasting the film with harsh criticism: "Regardless of whether or not my name appears on the screen among the credits, I'm not afraid that anybody will think I wrote this stuff. They'll know damn well I didn't. I shouldn't have minded in the least if you had produced a better script—believe me. I shouldn't. But if you wanted something written in skim milk, why on earth did you bother to come to me in the first place?"[2] Okay, so we'll count Raymond Chandler in the minority, as most people love the well-reviewed *Strangers on a Train*.

Robert Walker, the actor behind villain Bruno Antony, died shortly after the release of *Strangers on a Train* at only thirty-two years old. He had been struggling with psychiatric issues exacerbated by alcoholism. When a psychiatrist was called to his home, Walker was injected with a drug that it is believed he had an allergic reaction to. We consider Bruno Antony to be one of Hitchcock's best villains, so we have to wonder what other collaborations Walker and Hitchcock could've made if not for his tragic death.

Farley Granger, who played Guy Haines, the hero of *Strangers on a Train*, had previously appeared in Hitchcock's *Rope* (1948). He had a prolific career spanning films, television, and theater up until his seventies. Unlike many in the Golden Era of Hollywood, Granger did not feel the need to hide his sexuality from his peers. While moviegoers didn't know he was bisexual, those in Hollywood did, especially once he settled down with Robert Calhoun, his partner of more than forty years. The two cowrote the memoir *Include Me Out: My Life from Goldwyn to Broadway* (2008). Calhoun died in 2008 and Granger died, aged eighty-five, in 2011.

Another standout in *Strangers on a Train* is Hitchcock's daughter, who portrays Barbara Morton. Patricia Alma Hitchcock, the only child of Alfred and Alma, was born in London in 1928. From the time she could walk, she was on movie sets, including eating lunch with her dad on the train set of *The Lady Vanishes* (1938).

> According to [Hitchcock] O'Connell, her childhood was as normal and serene as it could possibly be as the daughter of one of the world's most famous moviemakers. Even after her parents moved to Hollywood in the late 1930s, so her father could make *Rebecca* for producer David O. Selznick, they lived frugally—Alma continued to do most of the family cooking—and held fast to family traditions. (Hitchcock, born and raised a Roman Catholic, drove his daughter to Mass every Sunday morning she was home.) Just as he always was a perfect gentleman on movie sets, he remained unruffled and soft-spoken at home. Rather than raise his voice when his daughter misbehaved, he would mournfully lecture: "Do you have any idea how much you've hurt your mother and me?"[3]

Patricia became involved in the family business, whether it was coaching young Edna May Wonacott in acting or acting herself in films like *Stage Fright* (1950) and *Psycho* (1960). Patricia's biggest film role was as Barbara Morton in *Strangers on a Train* when she was in her early twenties. She continued to work closely in her father's purview, including appearing in ten episodes of *Alfred Hitchcock Presents* (1955–1965), and was also known to tread the boards of Broadway. Shortly after Pat Hitchcock's turn in *Strangers on a Train*, she married trucking executive Joseph E. O'Connell. The pair had three daughters and were together until his death in 1994. Raising three girls became Hitchcock's focus, but she continued to come back in support of her parents' legacy up until her death in 2021 at age ninety-three. This includes coauthoring of a biography of her mother, *Alma Hitchcock: The Woman Behind the Man* (2003), and working with *Alfred Hitchcock's Mystery Magazine*. The magazine was established in 1956, inspired by Hitchcock's mastery of all things suspenseful. Each edition has several short stories, some of which were adapted for *Alfred Hitchcock Presents* back in the day. It is now the second-longest-running mystery magazine ever, with famous contributors of the past, like mystery authors Ed McBain and Donald Westlake. If you're like us and want more stories with the intrigue of Hitchcock's films, we suggest a subscription to the print or digital *Alfred Hitchcock Mystery Magazine*.

The climax of the film is not only one of cinema's most suspenseful but also one of the most technically daring. Guy chases Bruno onto a carousel, which starts spinning wildly. Of course, despite the risk of death there is comic relief, like the little boy (Louis Lettieri) who wants the carousel to go even faster. Guy and Bruno start punching each other, while a carnival worker (Harry Hines) attempts to slow down the out-of-control ride. Hitchcock spoke of this sequence to François Truffaut:

Alfred Hitchcock: My hands still sweat when I think of that scene today. You know, that little man actually crawled under that spinning carousel. If he'd raised his head by an inch, he'd have been killed. I'll never do anything like that again.

Francois Truffaut: But when the carousel breaks down—

Alfred Hitchcock: That was a miniature blown up on a big screen. The big difficulty with that scene was that the screen had to be angled differently for each shot. We had to move the projector every time the angle changed because many of the shots of the merry-go-round were low camera setups. We spent a lot of time setting the screen in line with the camera lens. Anyway, for the carousel breakdown, we used a miniature blown up on a big screen and we put live people in front of the screen.[4]

> Did you catch Alfred Hitchcock's cameo? He's hard to miss! Naturally, he is seen getting on the titular train. He's the one struggling to load the huge cello case onto the train; it's nearly as big as him.

A Diamond cab pulls up in the opening scene of *Strangers on a Train*, and it got us wondering how long taxi and ride-share services have existed. Before cars existed, a horse-drawn carriage was the most common form of getting a ride. Hackney carriages began in Paris and London in the early seventeenth century. These progressed to become hansom cabs, two-wheeled carriages that could be pulled by a single horse. By the end of the nineteenth century, electric battery–powered taxis were available, and in 1898 taxicabs in Paris were equipped with the first meters.[5] Now, taxis and ride-share services are a convenience we come to expect. CNBC reported that in 2023 there were roughly thirteen thousand taxis in New York City competing with eighty thousand ride-share drivers.[6]

Close-up shots show men walking in the film. What can we tell about a person by their gait? Many gait abnormalities can give clues about a person's health conditions. An antalgic gait results from pain, and you'll see someone limping or avoiding putting pressure on a certain part of the foot. Those with Parkinson's disease have a gait consisting of short, fast steps to maintain their center of gravity. People who have weakness of the muscles near their hips and gluteus area have a lurching gait that causes a slow, long stride.[7] Most people will experience a gait abnormality as they age or if afflicted with a condition.

As we rewatched these Alfred Hitchcock classics, we began to notice just how many characters were smoking! It's strange that in 2024 we don't see smoking as often on screen unless it is a period piece. Smoking was once considered healthy, and eventually, it began to be seen as a health risk. In the 1930s and '40s, doctors appeared in ads for cigarette companies, as the link between smoking and lung cancer had yet to be discovered. It wasn't until 1964, when the U.S. surgeon general's report was released, that facts were revealed about the dangers of smoking. Over thirty years later, the tobacco companies' research committee was forced to disband to stop the false narrative about cigarettes.[8]

As seen in the film, some people are natural conversationalists who can start talking to anybody and be able to keep the flow of conversation going. Sometimes, asking too many personal questions can put a stop to the conversation. This is explained by a theory called the Johari window, developed by psychologists Joseph Luft and Harrington Ingham in 1955 and still taught today in communications courses. (They combined their first names to create "Johari.") The Johari window contains four quadrants: the open self, the hidden self, the blind self, and the unknown self. The open area contains everything you would feel comfortable sharing with people. It could include your name, your occupation, your favorite restaurant, and other basic information. The hidden area is full of things you know about yourself but don't share with others. This could be your dreams, hopes, fears, things you're embarrassed about, or anything else you feel is too personal to tell just anyone. The blind self contains things other people know about you that you aren't aware of. These could be "tells," like in poker. Maybe you blink a lot when you're angry, or you get quiet when you're hungry. Unless someone informs you about your hidden self, you will be unaware. Finally, the unknown self are the things you and others don't know. Here lie the traits or reactions that don't pertain to you. Typically, the closer we get with someone, we'll reveal our hidden self, and they will reciprocate. When people jump ahead in conversation and self-disclose too much too soon, or ask invasive questions, it can make us feel uncomfortable.

Bruno has a theory that "you should do everything before you die." When did adventuring and thrill-seeking start to be documented? Sensation-seeking may have always existed, but it wasn't studied or named until the 1970s. Marvin Zuckerman, a psychology professor from the University of Delaware, created a personality test that assessed people's preference for stimulation. His research found that "high sensation seekers tend to seek high levels of stimulation in their daily lives" with males scoring higher than females.[9] This rush of adrenaline is only sought after by humans; no other animal seems to chase the high of thrill-seeking. Even though we enjoy new experiences, you probably won't find us skydiving or bungee jumping anytime soon.

Bruno puts himself to sleep at night figuring out the "perfect murder." What are some less creepy ways to fall asleep according to experts? The military method, developed during World War II, helped soldiers fall asleep in about two minutes. After six weeks of practice, the method had a 96 percent success rate. The method is as follows:

1. Relax your entire face, including the muscles inside your mouth.
2. Drop your shoulders to release the tension, and let your hands drop to the side of your body.
3. Exhale, relaxing your chest.
4. Relax your legs, thighs, and calves.
5. Clear your mind for ten seconds by imagining a relaxing scene.
6. If this doesn't work, try repeating "don't think" mentally for ten seconds.
7. Within ten seconds, you should fall asleep![10]

Other tips include progressive muscle relaxation, breathing exercises, and listening to guided sleep meditation.

The mama's-boy trope is used in *Strangers on a Train*. What is the psychology behind it? The term *mama's boy* was an insult, insinuating that a man was weak, but research shows that males who have "strong relationships with their mothers are mentally healthier, more empathetic, and have better relationships with women."[11] Being a mama's boy can become an issue if there are boundary problems, dependence, and unrealistic expectations of time or duties.

Bruno's mother paints a picture that she believes represents St. Francis, but it's a terrifying depiction of a monster. Bruno recognizes it as his father. This reminded us of the Rorschach test in which subjects' perceptions of inkblots are interpreted and analyzed by psychologists. Versions of the test have been used since the time of Leonardo da Vinci and are still used today in the United States. A survey found that 77 percent of clinical psychologists utilize the Rorschach test in assessment of patients.[12] It doesn't take a clinical psychologist to interpret why Bruno believes his mother's painting is of his father.

We see the characters in *Strangers on a Train* playing carnival games. What is the history of this pastime? Carnival games have existed since the Renaissance and became popular in the United States in the 1800s. The Chicago World's Fair in 1893 brought the carnival to the attention of the general public, and standard games, which we still play today, were established. Prizes weren't always a part of these games but are standard now.[13]

Guy needs an eyewitness to testify that he was on the train to Washington. He tries to get Professor Collins (John Brown), whom he spoke to, to vouch for him. The professor was drunk, though, and didn't remember Guy. How does alcohol affect our memory? Alcohol interferes with the ability to form new long-term memories and directly correlates to the amount of alcohol consumed. Becoming blackout drunk is common among binge drinkers, like college students, and induces episodes of amnesia.[14] No wonder the professor didn't remember Guy!

At the tennis match is an incredible shot of everyone turning their heads to follow the ball. It's a realistic and beautiful depiction of what spectators do when watching the sport—except one person, Bruno, is staring straight ahead. It's unnerving! People refer to Bruno as a lunatic in the film. Where does the expression come from? "Lunatic is a term referring to a person who is seen as mentally ill, dangerous, foolish, or crazy, conditions once attributed to 'lunacy.' The word derives from *lunaticus* meaning 'of the moon' or 'moonstruck.'"[15] As we've written about in our research previously, there is no scientific proof that the moon, or a full moon specifically, influences people, but many medical professionals share anecdotal information about how their patients react during changes in the lunar cycle.

Bruno accidentally drops Guy's lighter down a storm drain and must retrieve it. The task is difficult, but he's able to squeeze his arm through. How does the human body cope with getting into small spaces compared to other animals? Although some people can contort to fit into tiny areas, it's nothing compared to octopuses. "The only part of an octopus's body that's hard and solid is its beak, so most octopuses can fit through a space around an inch in diameter!"[16] Other animals that can fit through tiny openings include mice, cats, squirrels, and raccoons. A raccoon can get through a space four inches wide, a cat only needs three inches to contort its body, and a squirrel can get through a space as small as one and a half inches![17] We will be sleeping with our windows and doors closed tightly from now on and won't be speaking to strangers on trains.

The carousel scene in *Strangers on a Train* has epic fight choreography. "Minor or severe, around thirty thousand accidents in theme parks–related injuries are reported every year. From minor incidents like trip-and-fall accidents to more severe incidents involving ride malfunctions, these theme parks accidents can result in injuries or, in rare cases, fatalities."[18]

CHAPTER TEN

Dial M for Murder

The year 1953 became a banner one in the life of Alfred Hitchcock. He signed a deal with Paramount Studios that gave him even more control over his filmmaking team, and it contributed to his already growing finances. But he wasn't known to live an opulent lifestyle: "The Hitchcocks didn't live in a palace, and they never indulged in extravagant habits. Food, wine, and travel were luxuries the couple shared with friends and relatives."[1] They would pay for family to visit from the United Kingdom, although Hitchcock's mother refused to make the trip overseas. The director was generous when it was needed, like when his widowed sister-in-law entered a sanatorium and he set up a system to pay all her bills.

It seemed that for both Alfred and Alma, the work of filmmaking was what truly made them happy. They did add a wine cellar to their home on Bellagio Road in Los Angeles, as wine was a hobby of Hitchcock's. He was proud of his collection of vintages. His other indulgence was art—original works by Pablo Picasso, Amedeo Modigliani, and many others graced the walls of his Californian home. Biographer and film scholar Bill Krohn has studied Hitchcock's impressive art collection, writing on it extensively. If you're curious, the Hitchcock's former home is still standing, having been renovated after the director's death in the early 1980s. Most recently, it sold in 2023 for $8.8 million.

As I (Meg) was watching *Dial M for Murder,* I was reflecting on the setting, imagining that it would be perfectly suited as a staged drama. The vast majority of the action takes place in the apartment of Tony and Margot Wendice (Ray Miland and Grace Kelly). So, it wasn't a complete shock when I delved into the research to find that the film was based on a play. Written by Frederick Knott, the play had been successful in Britain; its launch had been televised on BBC in 1952. Knott wrote the screenplay for the film version as well. He is also known for writing the play *Wait Until Dark* (1966) which was adapted to the screen the following year by other screenwriters. In the *Wait Until Dark* (1967) film version, Audrey Hepburn depicts a blind woman embroiled in a crime plot. She was nominated for an Academy Award, and I know it is the Hepburn movie that I remember most vividly watching in my own childhood. It's terrifying.

Just as frightening is *Dial M for Murder*, in which Tony plots to kill his philandering wife, Margot. He blackmails a former friend, Charles Swann (Anthony Dawson), to do the deed while he is out benefitting from an alibi. The audience watches in hushed anticipation to see how this murder will play out.

Well, thankfully for the play attendees and the filmgoers, the murder does not go according to Tony's devious plan, and we get to watch him improvise when Margot kills Charles in self-defense. Oops! Tony has to think on his feet, which means he frames Margot to look like she killed Charles after him luring him there. The plot is complicated, hinging on different forms of evidence that are not always easy to follow (so many keys!) but is satisfying in its suspense.

The true star of *Dial M for Murder* is Grace Kelly, who, if you ask us, belongs on the silver screen. Her beauty is legendary, yes, but her acting is first rate. She embodies Margot with a subtle empathy, making us not judge the fact that she has a hidden boyfriend (Robert Cummings) for a second. In fact, it is quite obvious from the onset that Tony is the villain and that Margot is trapped in a terrible situation—one even more drastic in the 1950s. Kelly is known as one of Hitchcock's most prominent leading ladies (she also starred in *Rear Window* and *To Catch a Thief*), until she left Hollywood and became a princess! Just like out of a fairytale, Kelly wed Prince Rainier III of Monaco in 1956. Aside from a few documentary projects, she wasn't involved in Hollywood again, devoting her life to serving Monaco and to raising her royal family.

Hitchcock was known to have a penchant for classy, blonde female leads, women he was said to demand excellence from. This included him coaching Kelly in changing the very tone of her voice to play Margot Wendice. Some women (and men) bristled at Hitchcock's control over a film, but like Ingrid Bergman, Grace Kelly seemed to enjoy the director's nature. They both were fond of puns, joke-telling, and even practical jokes—though Kelly later recounted that it wasn't all fun and games on the set of *Dial M for Murder*:

> Hitchcock and Kelly did have a number of combative moments, however, which proved a little embarrassing for her on set. Kelly told Donald Spoto, "We were blocking a scene, and I was standing there, a little bewildered. Then I heard a voice calling me, 'Miss Kelly, what do you think you are doing?' I called back to him, 'I'm trying to figure out where Margot would glance, and where she would go at this moment.' And Hitch said, 'Well, Miss Kelly, if you had read your script properly, you would know that she is to look in this direction and go over here. Don't you ever read stage directions?' So, I was called down on that."[2]

On the flip side, "there was one instance where he conceded to Kelly's ideas on how her character would act. He had originally planned for Margot to put on a robe upon getting out of bed to answer the phone, but Kelly argued that no woman would take the time to do that if she were home alone. Hitchcock agreed to do it her way, and it stayed in the film that Margot answered the potentially fatal phone call in a mere slip."[3]

In the early 1950s, studios and movie theaters had a new competitor stealing their audience: television. With the advent of the networks came a decline in the amount of people paying for tickets at theaters. As modern moviegoers, we can recognize not only how tempting it is to stay on our couch and watch a quality show but also how far films have gone to entice us back into the cineplex. Every so often, a new batch of three-dimensional movies are released. As horror fans, we've come to enjoy the occasional 3D flick, from *House of Wax* (1953) to the remake of *My Bloody Valentine* (2009). *House of Wax* was the first horror film shot in 3D and the first 3D genre film with color. Its star, Vincent Price, became known as the "King of 3D," as he starred in many, especially the scary kind! We were lucky to have our friend Nate invite us over to watch his copy of the 3D version of *House of Wax*, cardboard glasses and all. It was a truly resplendent horror experience.

The first 3D movie was *The Power of Love* from 1922. In 2024, 3D technology exists for use in homes and movie theatres.

The period from 1952 through 1954 marked the first explosion of 3D in cinematic history, which put pressure on Hitchcock to shoot *Dial M for Murder* for 3D:

> It was filmed in the summer of 1953, when 3D was at its peak. The studio, however, delayed its release [until] Frederick Knott's play concluded its run.
>
> In May 1954, when *Dial M for Murder* was finally released, audiences and studios had lost interest in 3D. The film eventually opened in 2D and most audiences saw the film in its flat format. Alfred Hitchcock would later comment on 3D by saying, "It's a nine-day wonder, and I came in on the ninth day."[4]

It's odd to imagine *Dial M for Murder* as a 3D picture because it is one of Hitchcock's quieter pictures—suspenseful, of course, but all relegated to the Wendice's apartment. No Jason Voorhees was slinging a machete at the audience, nor was a thrilling car chase headed our way. We imagine, though, the scene in which Margot is strangled by Charles and she eventually stabs him in the back with scissors would be particularly effective in three dimensions.

We're impressed with Alfred Hitchcock's cameo in *Dial M for Murder*, as there were few opportunities for him to make an appearance in a film with minimal characters and settings. Did you spot him? When Tony shows Charles a photograph of them at a college event, which hangs on his wall, there is Hitchcock joining them at the table!

We see a character toss salt over the left shoulder in the film. When did this practice originate, and what does it mean? In Western culture, spilling salt is considered bad luck, so tossing it over your shoulder will reverse it. Why the left shoulder? According to some Christian beliefs, the devil hangs over our left side, so throwing salt in his face will prevent bad things from happening to us. Christians aren't the only ones to have superstitions about salt. Ancient Egyptians, Greeks, and Romans invoked gods with salt;[5] additionally, Wiccans believe salt cleanses an area of harmful, negative energy.[6]

In 2023 it was adapted again for the stage by Jeffrey Hatcher to portray the lovers as lesbians. This fresh take on the story resonates with today's audiences, as queer topics are at the forefront of people's minds. What is the science behind sexual orientation? In 2019 a study was published in *Science* that suggested human sexuality cannot be pinned down to just one factor. By studying nearly half a million people's sexual orientation, genetics, and backgrounds, scientists discovered that biology, psychology, and life experiences all played a role in how people identify.[7] Whereas biology alone can't predict a person's preferences for whom they are attracted to, the American Psychological Association recognizes that most people feel they have little or no choice in the matter: "Sexual orientation is distinct from other components of sex and gender, including biological sex (the anatomical, physiological, and genetic characteristics associated with being male or female), gender identity (the psychological sense of being male or female), and social gender role (the cultural norms that define feminine and masculine behavior)."[8] Due to the era in which Alfred Hitchcock lived, any queer representation in his films wasn't overt but instead became queer-coded through subtext.

To better understand the relevance and scope of Alfred Hitchcock's filmmaking, we spoke to Mike Scholtz. He is a documentary filmmaker who lives in the woods of northern Minnesota, where most of the stories he tells originate:

Kelly: Describe what your experience is with film and television.

Mike Scholtz: The first time I can remember actively thinking about how films were made was thanks to *Mister Rogers' Neighborhood* [1968–2001]. Mister Rogers would often use a screen on his wall called "Picture Picture" to show these short, experimental documentaries about how crayons or applesauce or graham crackers were made. At first, I was fascinated just to go behind the scenes in these factories. But then at some point, I started thinking about the way these videos were edited and scored, always, with light jazzy piano. Not long after that, I caught a behind-the-scenes documentary about the making of *Star Wars* [1977], and I was hooked on the idea of telling stories visually.

Meg: What was your first experience with Hitchcock? Do you have a memory of seeing one of his films for the first time?

Mike Scholtz: For someone obsessed with digging behind the scenes on how stories are told, [I think] Alfred Hitchcock is kind of the perfect filmmaker to discover at a young age. So many of his films are formal experiments in storytelling. *Lifeboat* [1944] takes place in one tiny location. *Rope* [1948] appears to be filmed without a single edit. *Rear Window* also takes place in a single location, except it's huge. *Rear Window* was actually the first Hitchcock I ever saw. I was home with the chicken pox, the same week that a new low-powered UHF station debuted in the Fargo-Moorhead TV market. For the most part, they aired hours and hours of *Green Acres* [1965–1971] reruns. But on Sunday night at 8:00 p.m., they finally played something worth watching: *Rear Window*. And that was my first Hitchcock film.

Meg: What's your educational background in regard to film?

Mike Scholtz: Somehow, the stars aligned to dump Alfred Hitchcock films in my lap. I went to college in my hometown because Moorhead State University had a burgeoning film-history program with an entire class on Alfred Hitchcock's films. They also offered classes on Charlie Chaplin, Stanley Kubrick, and Universal's monsters. Most other colleges and universities that I could actually afford offered nothing more than a single Film History class.

Kelly: What a great variety of classes!

Mike Scholtz: After I graduated from college, I got a job making industrial videos, where I got to tour and film the kinds of factories

Mister Rogers once featured on his TV show. Then I worked for a few more years at an advertising agency, where, I insisted, I was making thirty-second experimental films. But they were just TV commercials. All along the way, I was producing my own documentaries. *The Angela Murray Gibson Experience* [1997] was about a woman from tiny Casselton, North Dakota, who opened her own studio and produced silent films. *Wild Bill's Run* [2012] was an Arctic crime caper about a snowmobiling criminal. *Wicker Kittens* [2014] was about competitive jigsaw puzzling. Finally, about five years ago, I started making documentaries full-time. Now I do most of my work for a couple of public television stations in Minnesota. Thanks to Mister Rogers, I'm a huge fan of public TV.

"Public Broadcasting Service (PBS) was founded in 1969, at which time it took over many of the functions of its predecessor, National Educational Television (NET). . . . Since its founding . . . PBS has grown to include 354 stations which cover all fifty states."[9]

Kelly: You shoot and edit for a living. What stands out to you specifically in Hitchcock's films that sets him apart?

Mike Scholtz: As someone who loves documentaries, and specifically behind-the-scenes documentaries, I just love the way Alfred Hitchcock puts himself out there for our amusement. He famously inserts a cameo of himself in every one of his films. He introduces every episode of his '50s TV show *Alfred Hitchcock Presents*. He likes to shoot trailers for his movies that show him on set. Other directors want you to be swept up in the story. They don't want to risk distracting you. But Hitchcock's basically doing everything he can to remind you that these movies are make-believe.

Meg: That's a great point!

Mike Scholtz: I think that extends to his love for showy cinematography. In *Notorious* [1946], his camera zooms in from the ceiling to a key in the palm of a hand. In *Strangers on a Train*, he shows us a murder reflected in the victim's dropped eyeglasses. In *Vertigo*, he moves the camera back while zooming in to approximate the feeling of vertigo. This is all very showy, hey-look-at-me stuff. But it's also the kind of stuff that makes people want to be filmmakers. Or, at least, it makes them want to be film buffs and learn how those shots were made. I think that's what makes his filmography such a gateway drug to the rest of film history.

Meg: Finally, what is your favorite Alfred Hitchcock film?

Mike Scholtz: I've been thinking about this a lot. I've kind of talked myself into choosing *Rear Window* as my favorite Hitchcock, mostly because it was my first. It was also screenwriter John Michael Hayes's first film with Hitchcock. And those two made such an unbeatable team. All four of their films are jam-packed with style and wit and cool. I literally think about the set they built for *Rear Window* every day. (I probably think about the characters every other day or so.) But at the risk of repeating myself, just for fun, I do love to tell people that *To Catch a Thief* is Hitchcock's best work. And half the time I actually believe it. It's easily his funniest film. The dialogue is so sparkly! The costumes are so scrumptious! I think it comes down to my mood. The protagonist of *Rear Window* is played by Jimmy Stewart. He's an everyman caught up in a dangerous situation beyond his control. But the protagonist of *To Catch a Thief* is Cary Grant. He's the coolest person who ever lived caught up in a dangerous situation he's going to handle just fine. Jimmy Stewart is Charlie Brown. Cary Grant is Snoopy. And who do I feel more like on any particular day?

Thank you, Mike, for getting us to think about these characters in a new way! And speaking of *Rear Window*, read on. (It's next!)

CHAPTER ELEVEN

Rear Window

Unlike the simple, contained apartment setting in *Dial M for Murder*, the set for *Rear Window* is a technical marvel. Made to look like a block in Greenwich Village, New York, the entire structure of multiple apartment buildings, street in the background, and a courtyard were built on a Paramount studio in Hollywood:

> According to Steven Jacobs' *The Wrong House: The Architecture of Alfred Hitchcock* (2014), for months the director and set designers Hal Pereira and Joseph MacMillian Johnson did nothing but plan the design of what was to become the largest set ever built at Paramount. Hitchcock himself superintended the huge and complex construction that took six weeks to set up. The entire set was fit with a sophisticated drainage system for the rain scene and with an ingenious wiring mechanism for the highly complex lighting of day and night scenes in both the exterior of the courtyard and the interior of the apartments. Of the thirty-one apartments, twelve were fully furnished on a massive set that was 38 feet wide, 185 feet long and 40 feet high—with the soundstage floor removed so that the courtyard level could be built into the basement.[1]

Rear Window is a prime example of Hitchcock's ability to trap and keep us in an agitated state of pulse-pounding suspense. Like Jeff, we are made inert, unable to help as he watches Lisa confront Lars Thorwald across the street. We empathize with his frantic desperation, as we, like him, have no ability to change the course of events. We are left to watch from the shadows alongside him, as Hitchcock holds us back, rather than moving the camera closer to the action. Even when Thorwald comes to Jeff's apartment to kill him, the action right before us, we are again struck helpless as we watch disabled Jeff battle a killer with only a flashbulb. With every advance of Thorwald, the blinding flashbulb disorients the audience, too, again enhancing the suspense in a manner only Hitchcock is capable.

> Manufactured flashbulbs used for photography were first produced commercially in Germany in 1929 and could only be used once.[2]

The source material for *Rear Window* is the short story "It Had to Be Murder" (1942) by Cornell Woolrich. We read a PDF copy of the story and were surprised to see how differently the film turned out. In "It Had to Be Murder" there is no mention of Jeff being a photographer. No Lisa. No Nurse Stella. There is little mention of the other neighbors, and, if we're being honest (and we *always* are), the literary Jeff is rather unlikable. He doesn't have an ounce of James Stewart's charm. Story Jeff's abhorrent behavior is most obvious in the racist treatment of his Black servant, Sam. Both Jeffries and the author, Woolrich, mistreat Sam. *Cringe.* We're glad screenwriter John Michael Hayes scrapped the insensitive portrayal of Sam, instead giving us the dynamic Lisa and hilarious Stella as Jeff's partners in crime detection.

Some things are similar, like the involvement of a NYPD detective, an old friend of Jeff's, and, of course, the murder across the street. While we think the movie is infinitely superior, the story is imbued with a familiar sense of dread and paranoia: "Watching him across the night, I speculated: why doesn't he get out? If I'm right about him, and I am, why does he stick around after it? That brought its own answer: because he doesn't know anyone's on to him yet. He doesn't think there's any hurry. To go too soon, right after she has, would be more dangerous than to stay awhile."[3]

American author Cornell Woolrich is known to have transitioned from writing a spate of "jazz novels" in the 1920s, to his prolific, and more successful, turn in crime fiction up through the 1980s. Numerous movies have been adapted from his stories and novels, including the Shirley MacLaine–starring *Mrs. Winterbourne* (1996) and even *The Bride Wore Black* (1968) directed by Hitchcock's friend and author of *Hitchcock/Truffaut,* François Truffaut. The French film was inspired by Woolrich's 1940 novel of the same name, written under one of his pseudonyms: William Irish.

As we're on the topic of adaptations, we were surprised to discover that Woolrich was inspired by another writer's work to write "It Had to Be Murder." British author H. G. Wells, who wrote in virtually every genre, published the story "Through a Window" (1895), about a man with a broken leg who becomes obsessed with watching his neighbors. Whereas Wells is best known for his contribution to science fiction, writing genre classics like *The Time Machine* (1895) and *The Invisible Man* (1897), critics argue that he was also relevant in the literary world of metaphor and theme. "Wells's story is about the ways in which reality encroaches upon us: we cannot be passive observers, viewing the world outside our windows as mere entertainment. Such a detached attitude treats the world outside as somehow unreal, fabricated or constructed for our amusement, rather than what it is: the genuine day-to-day reality of other people's lives."[4]

Though Hitchcock's film is rooted in suspense and intrigue, we think H. G. Wells's warning of the moral downside of watching our neighbors comes through via Woolrich's story and John Michael Hayes's script. If you're curious to learn more about Hayes, who also wrote Hitchcock films *To Catch a Thief, The Trouble with Harry* (1955), and *The Man Who Knew Too Much* (1956), we recommend that you pick up a copy of *Writing with Hitchcock: The Collaboration of Alfred Hitchcock and John Michael Hayes* (2011) by Steven DeRosa. The author delves into the successful relationship of Hitchcock with the young writer, a vet of World War II.

Knowing his friend well, François Truffaut recognized why *Rear Window* appealed to Hitchcock:

Truffaut: I imagine that the story appealed to you primarily because it represented a technical challenge: a whole film from the viewpoint of one man, and embodied in a single, large set.

Hitchcock: Absolutely. It was a possibility of doing a purely cinematic film. You have an immobilized man looking out. That's one part of the film. The second part shows what he sees, and the third part shows how he reacts. This is actually the purest expression of a cinematic idea. Pudovkin dealt with this, as you know. In one of his books on the art of montage, he describes an experiment by his teacher, Kuleshov. You see a close-up of Russian actor Ivan Mosjoukine. This is immediately followed by a shot of a dead baby. Back to Mosjoukine again and you read compassion on the face. Then you take away the dead baby and you show a plate of soup, and now, when you get back to Mosjoukine, he looks hungry. Yet, in both cases, they used the same shot of the actor; his face was exactly the same. In the same way, let's take a close-up of Stewart looking out the window at a little dog that's been lowered in a basket. Back to Stewart, who has a kindly smile. But if in the place of the little dog you show a half-naked girl exercising in front of her open window, and you go back to smiling Stewart again, this time he's seen as a dirty old man![5]

The montage technique Hitch referred to is the Kuleshov effect. Vsevolod Pudovkin, studying under Lev Kuleshov, came to see how crucial sequence was in filmmaking. "At its core, the early Russian film theorists, like Pudovkin, believed that editing, the organization and placement of shots, was a means of expression that was unique to filmmaking—something that wasn't (and still isn't) done in literature, theater, paintings, or the plastic arts. The foundation of film art is editing.[6] As evidenced in his talk with Truffaut, Hitchcock was steeped in filmmaking theory from around the world, utilizing it in his groundbreaking work.

Alfred Hitchcock also mentioned the "dog in the basket," a clever trick in which an upstairs neighbor uses a basket to lower her tiny dog down to do his business in the courtyard. Unfortunately, (spoiler alert) the dog dies, murdered by Thorwald. Now we're furious! If you're anything like us, fictional dog deaths hit you harder than fictional human deaths. If a dead dog, cat, squirrel, or any other critter, is a definite no for you, we recommend you stop by the website www.doesthedogdie.com. Search a movie, TV show, or book, and there's a good chance the crowdsourced site will let you know if there is any pet carnage and what to expect. We searched *Rear Window*, and it is confirmed that the dog does indeed lose his life. That scoundrel Thorwald!

> Did you spot the director? Alfred Hitchcock makes his cameo in the apartment of the "songwriter" (Ross Bagdasarian) donned in coveralls, winding a clock.

Rear Window begins by showing the extreme heat in New York City, emphasized by sweaty brows, people sleeping on their fire escape, and a close-up of a thermometer reading nearly 100 degrees Fahrenheit. Although many of us can luxuriate in air conditioning on hot days, this has not always been the case. Cooling our homes in the modern sense wasn't invented until 1901, so generations of people have used techniques to escape the heat. In ancient Egypt, passive air-conditioning techniques were used to cool buildings, and this continued throughout the twentieth century. With experiments of chilling ice, using evaporation, and compressing and liquefying ammonia, scientists were able to develop various methods to condition air. With the advent of electricity, more households gained air conditioning, and by the late 1960s and early 1970s, it became standard in New York City. In 2023 over 90 percent of New Yorkers report having an air conditioner, but that number drops to 76 percent in poorer neighborhoods.[7] Besides comfort, why is it important to regulate temperature? Air conditioning can help prevent heat stroke, dehydration, and other health conditions relating to overheating and helps with filtration and disinfection of air. Former ways of cooling buildings are now being used in new architecture as growing concerns over environmental impact and climate change are at the forefront of people's minds.

L. B. Jefferies (James Stewart) is a photographer recovering from a broken leg in his apartment in *Rear Window*. He uses his camera and a long-angled lens to investigate the goings-on of his neighbors. What is the history of these devices? In 2024 we may take for granted that, through our phones, most of us always have a camera on us. Of course, this wasn't always the case. The camera obscura is considered the first camera; it was described as "relatively little gloomy rooms with light entering only through a small hole. As a result, the adjacent wall was cast with an inverted picture of the outside scene. This approach was used to see solar eclipses without harming one's eyes and, later, drawing assistance."[8] The first records of this device were attributed to Han Chinese scholar Mozi, who lived from 470 to 391 BCE. Throughout the following centuries, scholars, physicists, and architects studied and wrote about the camera obscura. Although a design for a handheld camera was created in 1685 by Johann Zahn, a physical camera didn't exist until 1816.[9]

Some may call Jefferies in *Rear Window* a peeping Tom. Psychology is behind voyeurism. The *Diagnostic and Statistical Manual* defines the condition as "the act of observing individuals, usually strangers, engaging in sexual activity, exhibitionism, or disrobing."[10] Colloquially, society has extended the definition to anyone who views the lives of others without their knowledge. This has been used to explain the popularity of reality television and docuseries in which audience members gain a perspective and insight into the lives of others that they usually wouldn't partake in. This certainly explains Jefferies's curiosities and can be seen as a way to pass time while he is confined in his domicile. Although this type of voyeurism can be seen as benign, some countries have enacted laws that make it a punishable offense.

Jefferies worries about the right time to marry his girlfriend Lisa (Grace Kelly) and is worried that he's not ready. This phenomenon is known as the social clock and refers to what we view is the right time and order for things to happen in our lives. For example, some couples believe they should get married before purchasing a house together. Others have a set age in their mind for the time they believe they should start having kids. The social clock varies from culture to culture and has changed greatly over the past several decades. The average age of couples getting married in 1954, when *Rear Window* was released, was their early twenties; in 2024, it's early thirties. This could be due to several factors, including the increase of women in the workforce, changes in societal pressure, and views of marriage in general.

While observing the neighbors, Lisa comments that she can tell the dancer across the way isn't in love with any of the men she is entertaining. Can we read others' nonverbal factors to understand their emotions? From a very young age, if we have emotional intelligence, we are able to read people's nonverbal behaviors to determine their mood. Lisa likely noticed lack of eye contact, proxemic distance, facial expressions, and tactile communication to infer the neighbor's feelings.

Jefferies is convinced that he and Lisa are too different to work out in the long run. Do opposites attract, and can these relationships work out? Whether it be with a significant other or a friend, there are two types of relationships: symmetrical and complementary. *Symmetrical* means you have much in common, while *complementary* refers to having opposite interests or skills. For example, you may be an excellent cook, and your partner can't make any dishes. You complement each other by maybe dividing the work of cooking and doing the dishes. Another example would be that you are shy, and your partner is outgoing. By going out together, you're able to navigate situations you wouldn't on your own. We both have spouses who have many traits opposite to ours, but we share similarities as well. Experts say these things we have in common, especially core beliefs, values, and hobbies, will keep us together longer.[11]

The trio of Stella (Thelma Ritter), Lisa, and Jefferies remind us of current-day crime solvers who take an interest in cases and help law enforcement discover clues. Although amateur sleuthing has been around since the nineteenth century, current day law enforcement sees potential risks in it. Tips from the public can help solve a case, but an abundance of social-media posts and various hypotheses can cloud or muddy an investigation. Others suggest that, with the public's help, more clues can be found and evidence, like video footage, can be viewed faster. There are organizations of regular people who are devoted to solving crimes, like Websleuths, the National Missing and Unidentified Persons System, and the Doe Network, and they are open to join.

Lisa offers many feminine perspectives on the disappearance of a neighbor's wife in *Rear Window* by stating a woman wouldn't leave her jewelry or favorite handbag behind if she was leaving. Does female intuition exist, and how does this play a role in investigations? It's imperative to first understand the gender differences in intuition and how it's viewed in society. As Maya Boyer wrote,

> When it comes to decision making involving instinct or intuition, men are more likely to be rewarded for this implicit knowledge whereas women's decisions and/or theories based on instinct are often met with doubt or resistance, despite the fact that some studies have found that women tend to have better social cognition and are more likely to pick up on nonverbal cues. Women are often taught to suppress these instincts in favor of other, supposedly more rational, ways of thinking and processing information, or in favor of compliance with social and gender norms. Intuition is a valid and essential aspect of life and human thought, and the dismissal of intuitive thought, specifically and especially with women, by others or by oneself is a dismissal of more holistic observation and thought and directly compromises women's personal safety and health.[12]

We loved seeing Grace Kelly scale a ladder and walls in a gown and high heels while simultaneously feeling the thrill and stress of her situation to look for the missing neighbor's wedding ring. Her intuition was spot on, and she is a hero of this story.

Jefferies uses the flash of his camera to distract and slow down Mr. Thorwald (Raymond Burr) in the dark apartment as he approaches him. How does bright light affect our eyes when they're adjusted to the dark? It takes human eyes an average of thirty to forty-five minutes to get used to low light, so a bright flash will start the process all over again. A camera flash would cause temporary blindness, and repeated exposure to a flash could harm the retinas of the eyes. Our pupils expand to let in more light and contract to keep out light. If you know you're going to be in a dark area, you can prepare your eyes by wearing red-tinted glasses for twenty to thirty minutes prior. Avoid looking at bright lights, and take time to let your eyes adjust naturally.[13]

> The 1998 remake of *Rear Window*, starring Christopher Reeve, plays out differently from the original with plot points and beat points (moments of change among characters) but is still a satisfying watch. It delves into representation of someone disabled on life support navigating life after an accident, and mobility is even more limited than for James Stewart's character.

We love how in the movie the songwriter and "Miss Lonelyhearts" (Judith Evelyn) get together in the end. A little romance in a Hitchcock film is always welcome. Interestingly, Bagdasarian, who played the songwriter, was a real songwriter and performer who you can thank for creating the animated empire that is *Alvin and the Chipmunks* (1958–present).

Rear Window has inspired numerous films over the years, including *Disturbia* (2007) and *The Woman in the Window* (2021).

CHAPTER TWELVE
Alfred Hitchcock Presents

If you're anything like us, you grew up watching reruns (or first runs) of *Alfred Hitchcock Presents.* It was a view into old Hollywood, into twisty, suspenseful stories that were quicker and perhaps easier to understand for kids than Hitchcock's more mature fare.

With the wild success of television, it was only a matter of time before Alfred Hitchcock would conquer the small screen. "CBS offered Hitchcock a state-of-the-art contract. He would lend his name to the series, serving as host and producer, directing a set number of episodes. His salary would be higher than he received for many of his feature films he had directed—reportedly $125,000 per episode. . . . *Alfred Hitchcock Presents* would boost the director's fame and wealth in ways he could have never guessed."[1] His first anthology series was inspired by his already impressive oeuvre, which was often based on literature and always chock-full of suspense. Unlike similar contemporary series like *The Twilight Zone* (1959–1964), *Alfred Hitchcock Presents* was more murder and intrigue, less ghosts and aliens. Patrick McGilligan explains more about the series:

> The tone of the series was at first decidedly English, with only a thin coating of Hollywood, and its British accent would persist throughout the run of the series. No other U.S. television show could claim quite the same pedigree, drawing on stories by a British Who's Who of authors, including H. G. Wells, A. A. Milne, Rebecca West, Julian Symons, V. S. Pritchett, Eric Ambler, and John Mortimer. Roald Dahl and Stanley Ellin were probably the most frequently adapted and more than once Hitchcock returned

to authors whose novels he had already filmed, Mrs. Belloc Lowndes (*The Lodger*), Ethel Lina White (*The Lady Vanishes*), and Selwyn Jepson (*Stage Fright*) among them.[2]

As a Roald Dahl fan, I (Meg) had to find out what episodes were based on his short stories. There were six, one of which, "Lamb to the Slaughter," was adapted by Dahl to the screen. This is one of the most well-known episodes of *Alfred Hitchcock Presents*, about a wife who kills her husband with a frozen leg of lamb—before serving it to the detectives for dinner. My childhood was steeped in Dahl's novels like *Matilda* (1988) and *The Witches* (1983), so I was curious to see what a mashup of Dahl and Hitchcock would be like. Unfortunately for me, when I went to the streaming services, it became clear that the Dahl-based episodes are no longer available to watch. Rumor on the internet is that it is a copyright issue. But don't fret! There are still seven seasons with over thirty episodes each to watch. And you may be able to find the missing episodes via YouTube if they haven't been removed.

Like many, we were first introduced to Alfred Hitchcock (the man, not the work) from his series. His dark humor comes to the forefront in his introductions and conclusions to each episode. In "Triggers in Leash," we watch as Hitchcock puts a revolver to his head and pulls the trigger several times. When nothing happens, he sighs and says, "See, that's why I don't like playing Russian roulette. I never win." This bleak absurdity is what makes his films, and his persona, so fun. He was also able to make fun of himself, like in the introduction for "Salvage" in which he pretends not to realize the camera is watching him as he berates cameramen as a reference to his rumored fastidious filmmaking.

> For each episode, Hitchcock filmed two openings and closings: one for American audiences, which made jokes about the sponsors, and one for European audiences, which made fun of Americans.[3]

Even though Hitchcock contributed a lighter mood to his series, the "plays," as he called them, are fraught with psychological terror. In its pilot episode "Revenge," directed by Hitchcock, Elsa (Vera Miles) is attacked by a stranger, sending her into a nearly catatonic state. When she points out the man responsible, husband Carl (Ralph Meeker) kills him. Yet, in the end, we find out that Elsa thinks *every* man in a gray suit is the malicious stranger. Oops. And yes, Vera Miles would play Lila Crane five years later. Lila is the sister of Marion (Janet Leigh), who uncovers the grisly truth behind Norman Bates in *Psycho* (1960).

The psychological tension ramps up in the second episode of the series, "Premonition," in which Kim (John Forsythe) forgets he killed his own father, a plot evocative of Hitchcock's *Spellbound*.

The series as a whole is a big reason why Alfred Hitchcock is so revered and remembered, while other talented directors don't get as much attention. The show boosted Hitchcock's profile, literally, as his sketched profile was the signature aesthetic of the show. The nine-stroke, simple drawing showcasing the director's round cheeks was created by Hitchcock himself. He used his talents from early title design to make a sketch that, to this day, is recognized worldwide as Hitchcock.

Another iconic touch to the show was the playful theme music. Like the sketch, it is synonymous with not just the series, but Alfred Hitchcock:

> For the music, [Hitchcock's] long-time collaborator, the composer Bernard Herrmann, suggested Charles Gounod's "Funeral March of a Marionette." It had been used for an American horror radio show "The Witch's Tale," which aired from 1931 to 1938. But Hitchcock reportedly remembered it from a 1927 film *Sunrise: A Song of Two Humans*. Gounod wrote the piece while residing in London, between 1871 and 1872. It was a satirical character piece that was intended to be a parody of the personality of a music critic, Henry Chorley. Gounod's English patron, Georgina Weldon, described Chorley as moving like a "stuffed red-haired monkey." For publication, however, Weldon titled the piece "Funeral March of a Marionette."[4]

As evidenced by her supporting role in *Strangers on a Train*, Hitchcock's daughter, Pat, was keen to continue acting in her dad's projects. She appeared in ten episodes of *Alfred Hitchcock Presents*, including "Into Thin Air," in which she plays Diana, a woman fearing for her own sanity amid the backdrop of the 1899 Paris World's Fair. In the introduction to the episode, Alfred Hitchcock relates its themes to his film *The Lady Vanishes*. When he concludes the episode, Pat's first in the series, Alfred, the proud dad, says, "I liked the main leading lady. She was quite good, don't you think?" In Meg's favorite episode, "The Older Sister," Pat Hitchcock plays the older sister, Margaret (whose real name was Emma), of accused murderer Lizzie Borden in this alternate look at true-crime history.

Alfred Hitchcock's domination of the TV didn't end with *Alfred Hitchcock Presents*. He also produced the *Alfred Hitchcock Hour*, premiering in 1962, which beefed up the stories from thirty minutes to an hour. This lasted three seasons and boasted such impressive writers as Ray Bradbury, Richard Matheson, and Robert Bloch, whose 1959 novel, *Psycho*, gave Alfred Hitchcock his biggest hit. This shift to hour-long episodes was important to Hitchcock, as he felt it allowed for better-paced suspense. For her piece for Slash Film, Debopriyaa Dutta spoke on the creative possibilities of the change:

This new, hour-long format was better suited to long-form storytelling, and the director was more than pleased to bluntly explain what this new format entailed in a newspaper interview in the same year: "It is decidedly true that The *Alfred Hitchcock Hour*, as the title rather plainly implies, will endure for sixty minutes each week, instead of thirty." Hitchcock also proceeded to criticize the show's sponsors for limiting precious runtime by clogging the broadcast with commercials, candidly stating that "despite the sponsors' infringement upon time, the one-hour period will allow [him and his collaborators] to tell full dramatic stories in natural narrative style, whereas the half-hour show permitted only short tales that led to a 'twist' ending." While twist endings have value, forcing such narrative pivots to accommodate a story within a time frame is hardly ever effective, let alone satisfying from a creator's perspective. The new format also allowed creators to focus on character depth, yielding more vibrant, believable stories that could still elicit a range of shifting tones. Hitchcock ended the interview by stating that suspense would "remain an active ingredient" in the rebranded series, and he delivered on this promise without compromise until the delightful, gripping anthology ran its course.[5]

Alfred Hitchcock Presents was revived in 1985 on NBC. Introductions by the director were colorized, while episodes were reshot or original episodes were written. The pilot aired in a movie of four episodes, featured both Tippi Hedren from *The Birds* and Kim Novak from *Vertigo*. It also starred Hedren's daughter, Melanie Griffith. The remake only lasted one year on NBC but enjoyed three more seasons on the USA Network.

Does Hitchcock make a cameo appearance in his show? Well, we don't think so. While he frames each episode with his hosting, there doesn't seem to be any episode in which he appears during the action. Are you a keen observer? Maybe you can prove us wrong. There *are* a few movies in which Hitchcock doesn't appear, too, like *Jamaica Inn* (1939) and *Sabotage* (1936).

"Breakdown," episode seven of season one of *Alfred Hitchcock Presents*, features a premise that has stuck with us since the first time we saw it. William Callew (Joseph Cotten) is unable to move or speak after a traffic accident but is completely conscious. We hear his inner monologue throughout the episode, and no one notices that he is still alive. He's able to move his pinky finger, but it's too late—he's brought to the city morgue and covered with a sheet. While this is a terrifying prospect, we wondered what the steps are that medical professionals go through to confirm death. The process should include observing "no palpable pulses, no heart sounds on auscultation (or asystole on ECG), no observed respiratory effort, no breath sounds on auscultation, and pupils dilated and not reactive to light."[6] At the end of the episode, a tear falls from William's eye as the coroner examines him, and he's saved.

"Conversation over a Corpse" premiered in 1956 and we loved the cold open of Hitchcock cooking, very scientifically, with lab equipment. (We don't recommend adding eggshells as he did to his recipe, though!) In this episode, one sister is persuading the other sister to poison their landlord. Cissie (Dorothy Stickney) and Joanna (Carmen Mathews) poison Mr. Brenner (Ray Collins), and he passes out.

> Joanna tests Mr. Brenner's arm in "Conversation over a Corpse" to see if he's really dead, and it is still limp. She may not know that rigor mortis doesn't set in for at least two hours in the facial muscles and several hours later in the limbs.[7]

As the sisters begin to move the body, Mr. Brenner wakes up, prompting the sisters to kill him again! They contemplate ways to finish the deed, including a gun, a knife, and a hammer. The gruesome scene plays out like comedy as they settle on a rolling pin from the kitchen. Neither sister can bring themselves to take the man's life, or so we think, as Cissie gets her revenge. Both her sister and Mr. Brenner are dead by the end.

In "Special Delivery," from 1959, Tom (Peter Lazer) receives a package containing spores that guarantee to grow mushrooms in less than twenty-four hours. Is this possible? Absolutely! Mushrooms are one of the fastest growing things on the planet, doubling size in just one day.[8] That is why so many food experts are trying to encourage mushroom consumption, because they are so renewable and easy to grow.

People are disappearing and Roger (Frank Maxwell) claims he has a bad feeling about what's going on. He encourages Bill (Stephen Dunne) to be more present and mindful. How can mindfulness help us in day-to-day life? Being mindful can help us slow down and make better decisions. If we practice mindfulness at a meal, we are less likely to overeat. When we're mindful during a disagreement or debate, we are less likely to say something we'll regret. Being present also helps us to slow down and appreciate moments throughout the day instead of rushing through and not noticing them. It's good advice, Roger!

Even though he was dishing out some great knowledge, Roger disappears. People speculate that perhaps he's having a midlife crisis since he's in his forties. How does aging affect us mentally? A midlife crisis is defined as "a period of emotional turmoil in middle age, around forty to sixty years old, characterized by a strong desire for change."[9] Many people reach a certain age and question if what they've done so far matters and if they are happy. It's also a time that people realize their own mortality and that can occur after a major life event, like divorce or a death in the family. If you think you are going through a midlife crisis, share your feelings with others, exercise more, and change your thinking. Being positive and grateful can adjust your outlook and help you get through a tough time.

Bill worries that the mushrooms are coming from outer space and could grow to a great size. What is the largest mushroom ever recorded? The largest living mushroom—and organism on Earth—is a fungus in Oregon. It covers 3.7 square miles, or over two thousand acres. Its type is *Armillaria ostoyae,* and its nickname is "The Humongous Fungus."[10] Although the mushroom grows relatively slowly (about three feet per year), it can take over and kill trees by feasting on their roots. This type of mushroom can be found in Europe, Asia, and North America. This terrifies us, and we won't be traveling to visit this site.

The episode ends implying that mushrooms are taking over the world, and this reminds us of one of our current favorite shows, *The Last of Us* (2023–present), in which a variant of the cordyceps takes over the planet. Is it possible for mushrooms or fungus to grow in or on humans? In our research, we came across a chilling story: "In 1950, a doctor treated a thirty-three-year-old man for fungal overgrowth of his toes. Upon isolating the fungus, the doctor discovered that his patient's foot infection was attributed not to any of the usual mold-producing suspects, but instead to a mushroom-forming species that commonly grows on trees. Since that shocking discovery seventy years ago, researchers have documented this wild fungus growing on and within other human bodies. To date, almost one hundred cases of infection and a few unexpected deaths have been reported."[11] The mushroom is called a Split Gill, and it grows all over the world on or close to dead trees or rotten wood. Although it is edible, we've decided to avoid this particular strain.

In 1960's "Insomnia," Charles (Dennis Weaver) suffers from lack of sleep. While his insomnia is caused by the memories of a fire that killed his wife, others experience insomnia for a variety of reasons. One is the recurrence of nightmares. How can you handle them if this happens to you? First, make sure you are relaxed and comfortable. Next, pick the recurring nightmare you'd like to work on. Write down every detail you remember, including sensory ones, and then change the ending. Create and write a new ending to the nightmare so that it turns into a positive. Throughout the day, read what you have written and imagine it in your mind. The nightmare should take on the new ending when you go to sleep after following these steps.[12]

Charles, unfortunately, perishes in a fire just like his late wife did, as he's finally able to sleep in the end.

"Summer Shade," from 1961, tells the tale of a couple who moves to Salem, Massachusetts, with their daughter. Salem, of course, is known for its history of witch trials, and strange things begin to happen. Kate (Susan Gordon) makes friends with someone imaginary. How many children have imaginary friends? According to a study in 2015, 65 percent of seven-year-old children had an imaginary friend at some point. Further research suggests that kids with imaginary friends "may develop language skills and retain knowledge faster than children without them."[13] Other benefits of having imaginary friends include improved problem-solving skills, improved management of emotions, and a tendency to be less shy. The father insists it's a girl thing. Do girls have more imaginary friends than boys? Yes, due to the trait of females seeking more social support than males.

Letty, the imaginary friend, gives Kate buzzard bones to "keep off the pox." What superstitions involve birds? If you've ever had a bird defecate on your head, like I (Kelly) did while walking around New York City, know that it's supposed to be good luck. Also know it's difficult to get out of your hair! Birds that accidentally fly into your home bring bad luck, while seeing an albatross as you travel on the water is a good omen.[14]

The minister in "Summer Shade" dismisses the old salves and tonics of the 1600s, but as we learned in our 2022 book, *The Science of Witchcraft*, plenty of natural medicines have always worked for treating illnesses throughout history. Read more about it if you're interested.

"The Kerry Blue" is an episode of *Alfred Hitchcock Presents* from 1962 that focuses on a man who loves his dog, Annie, so much he becomes obsessed when she passes away. Wouldn't we all? Before the dog's death, the wife appears to be jealous of Annie, claiming she eats better than the couple. What are the statistics regarding people and their pets in this era? In 2024 it's estimated that 66 percent of U.S. households own a pet, and many of the owners treat their pets like children.[15] You don't have to scroll long on TikTok to discover pets dressed in outfits, getting "pawdicures," facials, massages, and even chiropractic treatment. Why is this? Well, it could be because of the lowering birthrate in America. "Nearly eighty-five million American households have dogs. Only thirty-five million have children. There are about four million babies and six million puppies born in the U.S. every year."[16] As pet owners and parents, we fit into the category of having both—but admit we probably spoil our pets more.

Ned (Gene Evans) suspects his wife, Thelma (Carmen Mathews), buried the dog alive and ends up trying to murder her with an entire bottle of sleeping pills. Thankfully, she survives, but Ned perishes from a fall after seeing a dog in the backyard he believes is Annie.

Most of the stories in *Alfred Hitchcock Presents* end with the same moral: be honest, don't be a jerk, and don't be selfish! If you follow these rules, you should be just fine—at least in the Hitchcock universe.

CHAPTER THIRTEEN
Vertigo

Are you a fan of *Vertigo*? We're betting since you picked up this book, there's a good chance that you are mesmerized by the performances, the thrilling use of technicolor, and probably the harrowing ending. Perhaps not as widely known as *Psycho*, *Vertigo* is often touted as Hitchcock's best film by a smattering of critics. Yet, it didn't find its devoted audience right away, as Ty Burr recounted for the *Washington Post*:

> Once dismissed as a qualified misfire that narrowly broke even at the box office and won two "lesser" Oscars (for production design and sound mixing), Alfred Hitchcock's 1958 suspense thriller continues to blossom into fresh relevance with each new generation of film lovers. At the same time, this is the "Great Movie" that most daunts a casual viewer, especially since 2012, when *Vertigo* knocked *Citizen Kane* out of a half-century in the top spot on *Sight and Sound*'s once-a-decade poll of the best films of all time. That's a heavy load to carry for an odd, dreamlike movie that doesn't seem to care about meeting an audience halfway and that only becomes richer, sadder and more profound with multiple viewings.[1]

We agree that with repeated viewings the light humor of the film's beginning (James Stewart *is* charming after all) quickly disintegrates into a sad story punctuated with dread, and yes, that trademark Hitchcock suspense. That is especially true at the midway point, when Scottie, and the audience, believe Madeleine Elster (Kim Novak) dies from falling off a roof—which she kind of does and kind of doesn't. It's from that point onward that things really get strange, including Scottie's dream sequence that harkens back to the odd designs of Salvador Dalí in *Spellbound*'s significant dream on celluloid.

In a 2021 interview with *The Guardian*, Kim Novak spoke about her time working with Alfred Hitchcock, echoing the sentiment that parts of *Vertigo* didn't hold logic: "On set, Novak told Hitchcock there were bits she could make no sense of. 'I said: "I can't understand why you see Madeleine in the hotel window and then she disappears. How does she leave the hotel?" And (Hitchcock) said: "Ah, my dear, everything doesn't have to make sense in a mystery." '"[2]

Kim Novak is ninety-one years old. She gave up acting decades ago and now spends her time painting. In several articles she's made it clear that Alfred Hitchcock treated her with respect, but there were other men in Hollywood who did not. In a 2018 interview with the *Chicago Times* in honor of *Vertigo*'s sixtieth anniversary, in the wake of the Me Too movement, the actress shared some of her most trying times in the industry. She indicated that the dual nature of her role in the film, as well as Scottie's insistence her character change into another woman, was eerily similar to her own life:

> "I identify so very completely with the role because it was exactly what Harry Cohn and what Hollywood was trying to do to me, which was to make me over into something I was not," says Novak, referring to the iron-fisted Columbia Pictures founder who contracted her. "In the beginning, they hire you because of the way you look, obviously, and yet they try to change your lips, your mouth, your hair, every aspect of the way you look and the way you talk and the way you dress. So, it was constantly fighting to keep some aspect of yourself, trying to keep some of you. You feel: There must have been something in you that they liked, and yet they wanted to change you."[3]

Sixty years later, this astute assessment of being a woman in Hollywood, as well as how it mirrored how Scottie bullied Judy Barton into changing her appearance and manner until it ultimately killed her, makes Kim Novak more than just a Hitchcock Blonde in our eyes.

As with other Hitchcock films, *Vertigo*'s screenplay is based on a novel. This time, though, it's a translated book, *D'entre les morts* (1954), meaning "From Among the Dead," by French author duo Pierre Louis Boileau and Pierre Ayraud (a.k.a. Thomas Narcejac). Known under the simple moniker of Boileau-Narcejac, the authors were impressively prolific, writing one hundred books together. I guess this means we have a long way to go to be in their company. Only about ninety more books to go, Kelly!

The book is also known as *The Living Dead* in English and is considered one of the duo's most famous. They also cowrote *Celle qui n'était plus* (1952), or *She Who Was No More*, which was adapted into the French film, *Les diaboliques* (1955). The film proved popular around the globe, inspiring American remakes like *Diabolique* (1996), starring Sharon Stone. Mike D'Angelo wrote about it for the AV Club:

> According to various sources—none of them definitive, but the legend lives on—Alfred Hitchcock attempted to option Pierre Boileau and Thomas Narcejac's novel *She Who Was No More*, only to discover that French filmmaker Henri-Georges Clouzot had beaten him to it by a matter of hours. If that's true, it may have indirectly led to the greatest movie ever made (per the most recent *Sight & Sound* poll, anyway), as Hitch subsequently made a point of securing the rights to Boileau-Narcejac's *D'Entre Les Morts*, from which he created *Vertigo*. Still, it must have pained him a little to watch Clouzot's adaptation of the earlier novel become a cultural phenomenon. "An extraordinary new motion picture by Henri-Georges Clouzot, France's master of suspense" read the American ads, explicitly positioning *Diabolique* as Hitchcock with subtitles. That same campaign included a plea for viewers not to reveal the ending to their family and friends, along with a stern warning that nobody would be admitted to the theater after the movie had begun—ploys that Hitchcock would borrow a few years later for the release of *Psycho*. It's as if not being able to make this particular movie fueled his creative drive for much of the next decade.[4]

Regardless of however much of this is true, we're happy Hitchcock made *Vertigo*, as it has made an indelible mark on cinematic history. It's known to have influenced such auteurs as David Lynch, Brian De Palma, and, of course, Hitchcock's good friend François Truffaut.

If you're interested, you can walk your own self-guided *Vertigo* tour of San Francisco, taking in such sights as the Mission San Juan Batista, where the notorious movie deaths occur. But we should warn you: no actual tower was in the mission; it was added in with a little painted movie magic. Even better, if you're a devoted *Vertigo* fan, you can go on a guided two-hour tour usually given on Sunday afternoons by San Francisco City Guides for only a $20 suggested donation!

> Did you get dizzy trying to find Alfred Hitchcock in *Vertigo*? Yet again carrying another musical-instrument case (remember the oversized one in *Strangers on a Train*?), Hitchcock walks through Elster's shipyard with a bugle case in hand, wearing a sharp blue suit.

We spoke to Greg Carlson, a professor in the Communication Studies and Theatre Art Department at Concordia College in Moorhead, Minnesota, about his career and how Hitchcock has inspired him:

Meg: Tell us about how you developed an interest in filmmaking.

Greg Carlson: I have been a cinephile since I saw films like *Star Wars* and *Snow White and the Seven Dwarfs*. When I was in the third grade, I brought my program from a screening of the 1925 Lon Chaney version of *The Phantom of the Opera* to show-and-tell. I studied film and telecommunication as an undergraduate at Minnesota State University Moorhead (then known as Moorhead State University) under the mentorship of Ted M. Larson, who was a major collector of 16mm motion pictures. With Rusty Casselton, Ted actively sought rare silent films to restore, archive, and screen for the public.

Kelly: That's such important work!

Greg Carlson: Like many movie-obsessed colleagues and friends, I have a voracious appetite for films of all kinds. I like to travel to film festivals like Sundance, and I often bring students with me just as Ted Larson took students to events like CineFest in New York and Cinecon in Los Angeles.

Kelly: Was there a specific Hitchcock movie that sparked your interest? What is it about his filmmaking that still resonates with film scholars, students, and fans today?

Greg Carlson: The first Hitchcock movie I watched was *Psycho*, which I saw on the small television in the kitchen of the house where I grew up in Moorhead, Minnesota. I wasn't yet a teenager—I may have been about ten or eleven years old—and I had to beg my mom to let me see it. She was afraid that it might be too scary or too intense, and she was right. I had nightmares for several days. What scared me the most about *Psycho* wasn't the shower murder. It was the overhead shot of Mother attacking Arbogast on the stairs that played in my head on a loop.

Meg: That moment is terrifying.

Greg Carlson: "My mentor Ted Larson taught an undergraduate course on the films of Alfred Hitchcock. By the time I took the class, I had already seen *Rear Window, North by Northwest*, and a small handful of others. The class was a dream. Every Monday night, our group of enthusiastic scholars would spend more than three hours watching and discussing Hitchcock movies. We read essays on individual films and used the Donald Spoto book *The Art of Alfred Hitchcock* as one of our main texts. Since then, I have enjoyed the work of so many incredible academics who have devoted time and energy to writing about and understanding the work of Alfred Hitchcock. The list is too long to fully enumerate here, but some of my favorites include Modleski's *The Women Who Knew Too Much*, McGilligan's *A Life in Darkness and Light*, Rebello's *Alfred Hitchcock and the Making of Psycho*, Durgnat's *A Long Hard Look at "Psycho,"* White's *The Twelve Lives of Alfred Hitchcock*, and, of course, the *Hitchcock/Truffaut* transcriptions of the conversations between the two directors.

Kelly: One of our favorites!

Greg Carlson: *Vertigo* is my favorite Hitchcock movie, and I try to read anything I can find concerning the masterpiece. The Aulier book is a good place to start, as is the Barr monograph. I also recommend *The San Francisco of Alfred Hitchcock's Vertigo: Place, Pilgrimage, and Commemoration*, which was edited by Douglas Cunningham. I think that admirers of Hitchcock's work respond to that blend of restraint and ecstasy that accompanies his carefully cultivated persona. It reminds me so much of how we think and talk about David Lynch. Hitchcock loved playing the proper, suit-and-tie, English gentleman, pretending to be shocked by the ghoulish murders and wanton acts of violence he relished exploring on film.

Meg: Tell us about the *Vertigo* tour you got to experience.

Greg Carlson: Many years ago, I heard about Hitchcock fans tracing the locations used in *Vertigo*, and I filed the dream in the back of my mind for some future date. In the summer of 2022, my wife and I traveled to San Francisco, and we took Jesse Warr's incredible *Vertigo* tour, a guided odyssey to visit the places immortalized in the movie.

There are several websites devoted to finding the various *Vertigo* landmarks. They came in handy, along with a fresh rewatch of the movie and some supplemental books. I loved every minute of the tour and hope to return for an encore.

Thank you to Greg for the recommendations!

What causes the actual condition of vertigo and how can it be treated? Vertigo is the sensation of motion or spinning that is often described as dizziness. When experiencing vertigo, people feel as if they're spinning or that their surroundings are turning around them. The two types of vertigo are peripheral and central. Peripheral vertigo is caused by a problem in the inner ear that controls balance, which could come from swelling of the inner ear, an injury, certain medicines, or pressure on the vestibular nerve. Central vertigo is caused by a problem in the brain such as a blood-vessel disease, certain drug effects, multiple sclerosis, stroke, or tumors.[5] To treat vertigo, doctors may prescribe medicines to treat the symptoms, physical therapy to improve balance, or an Epley maneuver to place the head in various positions.

> The fear of heights is called acrophobia and affects between 3 percent and 6 percent of people. Exposure therapy has been proven to help with this phobia.[6]

Regarding fear of heights, which is prevalent in this film, we wondered how high a drop is fatal. Although a fall from as little as six feet can kill a person, some people have survived quite extraordinary falls. "In 1972, Vesna Vulovic, a cabin attendant, survived a 10,160m [33,000 feet] fall when the DC-9 she was in exploded over what is now the Czech Republic. [In 2004] a 102-year-old woman survived after toppling from her fourth-floor balcony in Turin."[7] Most people who fall are injured or die due to a fractured spine or damage to other vital organs. Landing on one's side has the best chance for survival.

Exposure therapy for various fears and phobias has been used since the 1950s. The idea behind this practice is to expose people to what they fear and avoid in order to help them overcome it. This could involve directly facing, imagining, or using virtual reality to confront a feared object, activity, or situation.[8] Scottie tries this technique at the beginning of the film by climbing on a stool, but is soon reminded of the rooftop where his colleague fell to his death and is overcome by anxiety.

Scottie believes Madeline is possessed by the spirit of Carlotta, a woman from the past. What is the psychological explanation of possession? According to the DSM-5, "In clinical psychiatry, trance and possession disorders are defined as 'states involving a temporary loss of the sense of personal identity and full awareness of the surroundings' and generally classed as a type of dissociative disorder. People alleged to be possessed by spirits sometimes exhibit symptoms similar to those associated with mental illnesses such as psychosis, catatonia, mania, Tourette's syndrome, epilepsy, schizophrenia, or dissociative identity disorder, including involuntary, uncensored behavior, and an extra-human, extra-social aspect to the individual's actions."[9]

Reincarnation is another possibility for Carlotta to be present. How many people believe in this phenomenon? A Pew Research Center study showed that 33 percent of Americans believe they will be reborn in another body after death[10] and a Gallup poll revealed that 65 percent more people believe in reincarnation now compared to twenty years ago.[11]

Hitchcock is known for casting blonde women in his films, and we had to wonder: do blondes have more fun? Being a natural blonde, I (Kelly) can only speak for myself, and I say yes, but studies show proof to back it up. Two separate studies had women go out with different colors of hair, either dyed hair or wearing wigs; when the women were blonde, they were approached more often than those with any other hair color.[12] This boost in self-confidence can translate to having more fun—or it could be annoying if you're just trying to vibe by yourself.

> Only 2 percent of people in the world are naturally blond, while in the country of Finland, 80 percent of the population is blond![13]

In *Vertigo,* Carlotta went mad from losing her child and died by suicide. Is the suicide tendency genetic? A 2021 study on genetics and suicide concluded that there is "a region in the genome on chromosome 7 containing DNA variations that increase the risk that a person will attempt suicide."[14] Worldwide, suicide accounts for almost eight hundred thousand deaths per year, so this discovery is important for treatment and prevention strategies to be researched.

Madeleine mentions that Scottie's questions are direct, and he says, "I don't mean to be rude." Depending on how we're raised, direct speech can come off as rude, whereas passive, or indirect, speech can be preferred. For example, if you say to someone, "It would be great if you could put the dishes away at some point before dinner," this would be indirect. It's almost a suggestion instead of an order. A direct way to say this is, "Put the dishes away right now." You can still be polite and say "please" and "thank you," but direct speech equals clearer communication, and the task is more likely to get done.

Scottie follows Madeleine in the film, and we assume she doesn't see him. Do we notice when we're being followed? We do!—at least, if we catch someone looking at us. Humans have what's called a "gaze detection system" and it's a sense that evolved to help us survive. "A direct gaze can signal dominance or a threat,"[15] so we naturally pay attention when this happens.

> In *Vertigo* John and Madeleine visit sequoia trees to wander about. "Giant sequoia are generally well able to protect themselves against their natural threats, allowing them to survive for thousands of years. They are too massive to be blown over in the wind, and their bark is thick and rich in tannins, which protect them against fire and insect damage."[16]

The lesson we took away from *Vertigo*? Don't change yourself for a man, or anyone for that matter. And be careful when you're on rooftops. Those things can kill you.

CHAPTER FOURTEEN

North by Northwest

In 1959 Alfred Hitchcock went back to his spy-thriller roots with his only MGM picture, *North by Northwest.* This was his fourth and final collaboration with Cary Grant after *Suspicion, Notorious,* and *To Catch a Thief.* Whereas *Vertigo* made a lackluster performance at the box office, *North by Northwest* was a huge hit for Hitchcock, paving the way for his next smash, *Psycho,* and also igniting a Hollywood trend of spy thrillers in the earlier 1960s.

The plot of the film harkens back to *The 39 Steps*, in which an innocent man is on the lam, dodging thugs, avoiding capture, and, of course, stumbling upon a beautiful and mysterious woman. In this case it's Eve Kendall (Eva Marie Saint). Oscar Award–winning Saint is a prolific actress with impressive credits. Her first film role was opposite Marlon Brando in *On the Waterfront* (1954), and she has continued to act for seven decades. She is one of the last living stars of the golden age of cinema. She will celebrate her 101st birthday on July 4, 2025.

> Aw, poor Alfie. He missed the bus! In the opening sequence of the film, Hitchcock is faced with the slamming doors of a bus. At least he's not carrying a heavy instrument this time!

Unlike many of Hitchcock's films, *North by Northwest* did not have source material like a novel or short story. In fact, the first seed of the idea came from the director himself. He had an image of Cary Grant hiding in Abraham Lincoln's nose—yep, that's how it all started. And the screenwriter, Ernest Lehman, who'd previously written *Somebody Up There Likes*

Me (1956) and *The King and I* (1956) was struck with inspiration. Sven Mikulec described the initial brainstorming for Cinephilia & Beyond:

> "I want to do a Hitchcock picture to end all Hitchcock pictures," allegedly announced screenwriter Ernest Lehman, after his friend, composer Bernard Herrmann, recommended him to Alfred Hitchcock. This inspired idea was born at the time when Lehman and the famed filmmaker were experiencing a sort of a writer's block—they were supposed to be making *The Wreck of the Mary Deare* for MGM, but when Lehman honestly stated he felt there was nothing he could do about advancing with the script, Hitchcock simply suggested they did [*sic*] something original. Directly out of this writer's block came *North by Northwest*, one of the most entertaining movies of the period and, perhaps, one of the most stylish chase thrillers of all time. . . .
>
> The film went through several title changes during development, being known as *The Man in Lincoln's Nose, Breathless, In a Northwesterly Direction*. . . . The final title was hardly Hitchcock and Lehman's first choice, but they simply failed to come up with an entry that satisfied them completely.[1]

Like all good stories, many elements were brought in to create a complicated and entertaining story. Several years before Hitchcock and Lehman began drafting *North by Northwest*, Hitchcock had chatted with a journalist named Otis L. Guernsey. Guernsey had told Hitchcock about an operation in World War II in which the British government had given a fake identity to a dead man in order to deceive the enemy. Lehman later confirmed this real-life inspiration of a not-so-real spy in interviews, saying that Hitchcock had been sitting on the idea for years, wondering how to use it. Athough the name of the operation wasn't given to Hitchcock, many believe it was Operation Mincemeat. This is when two members of British intelligence took the body of an unhoused man, Glyndwr Michael, who'd died from eating rat poison, and dressed him up, giving him personal items, as the fake Captain William Martin. On the body was planted correspondence that the Allies were going to attack Greece and Sardinia. This was successful, as they were able to invade and liberate Sicily, as they had planted, as Dame Agatha Christie would say, a red herring.

Perhaps the plot of *North by Northwest* harkens back to Hitchcock's earlier espionage films, yet MGM provided a budget that made for a truly different cinematic experience. Mount Rushmore had to be recreated on a sound stage, as there was no safe or sanctioned way for them to climb around on the real monument. Filming was done in South Dakota, in the parking lot and cafeteria of Mount Rushmore, but this after a profound push and pull between Hitchcock and the National Park Service even before he'd started filming *North by Northwest*:

> The filming of that scene, one of the most iconic in movie history, dominated *Variety*'s coverage of *North by Northwest*'s production. That was due in large part to a standoff between Hitchcock and the National Park Service and the Department of the Interior about his desire to film at the South Dakota monument. The director had long toyed with the idea of setting an action sequence on the Black Hills statue, reportedly telling associates, "I want to have one scene of a man hanging onto Lincoln's eyebrows. That's all the picture I have so far." He got his chance with *North by Northwest*, but only after the film's backer, Metro-Goldwyn-Mayer (MGM), promised government officials that no violent sequences would be shot "near the sculpture" or the talus slope. That prohibition extended to any mock-up of Mount Rushmore. As Todd David Epp writes in *Alfred Hitchcock's Expedient Exaggerations and the Filming of North by Northwest at Mount Rushmore*, the agreement was eventually revised to give Hitchcock permission to shoot a chase scene using a recreation of the monument provided that ". . . the presidents' faces be shown below the chin line in scenes involving live actors."[2]

Apparently, when the Park Service saw the chase sequence shot in the studio version of Mount Rushmore it was furious. They felt MGM had betrayed its word. The officials were worried that the constructed monument was so realistic that audiences would believe that if Roger Thornhill, played by Cary Grant, threw a thug over the side of the mountain, that would be in poor taste, or would be seen as consecrating sacred ground. Acting Secretary of the Interior Elmer F. Bennett threatened the president of MGM, saying that the National Park Service would be more scrutinizing of any filming in national parks moving forward. We can safely say we don't think Alfred Hitchcock was invited to any National Park galas after he filmed *North by Northwest*, but we're in awe of his commitment to getting Cary Grant into Abe Lincoln's nose.

If you're interested in learning more about the making of *North by Northwest*, we suggest you check out the 2000 documentary *Destination Hitchcock: The Making of "North by Northwest,"* which features actors Eva Marie Saint, Martin Landau, and Hitchcock's daughter, Pat.

Roger Thornhill, played by Cary Grant in the film, was forcibly intoxicated by men who wanted to get him drunk. This is considered a crime called involuntary intoxication, a situation in which someone is tricked or forced into consuming a substance like drugs or alcohol. This can also be used as a defense to a general-intent crime because it prevents the defendant from understanding the nature of his or her actions and differentiating between right and wrong.[3] Thornhill drives sporadically, nearly hitting several cars. He finally stops, after almost hitting someone on a bicycle, and is swiftly brought in by police for driving while intoxicated.

Even though Thornhill didn't drive drunk purposely, it's still a serious crime. According to the Centers for Disease Control, "Motor-vehicle crashes involving an alcohol-impaired driver kill over thirty people every day"[4] because alcohol impairs judgement, coordination, reaction time, sight, and concentration. An impaired driver may be given a breathalyzer test or a blood test to ascertain their level of intoxication. Blood tests are more accurate and are preferred since they are more likely to hold up in court.

Thornhill is being mistaken for someone named George Kaplan, who actually doesn't exist. Have there been cases of mistaken identity in which people were charged? One notable case in history involved a man in 1858 named William Armstrong, whose lawyer was none other than Abraham Lincoln. Lincoln proved Armstrong was innocent of murder by using a farmer's almanac to demonstrate that the position of the moon couldn't have provided sufficient light for the witness to see clearly.[5] A more recent case of mistaken identity took place in Minnesota in which a man, Trayvon Alexander, was mistakenly connected to his twin brother's criminal record. This created problems as he tried to apply for jobs. Thankfully, through the process of expungement, Alexander's records were cleared.[6]

When Cary Grant is shaving with a women's razor in the train station bathroom, we wondered about the history of hair removal for women. The first evidence of such comes from ancient Egypt. Both men and women shaved their entire bodies for hygienic reasons and used bronze razors, pumice stones, and creams made of resin and beeswax.[7] Throughout the centuries, hair removal went in and out of fashion, and it wasn't until the twentieth century that shaving products for women started to be marketed and produced. In 2021 the global shaving market size was valued at $4 billion and is expected to grow.[8]

> The most iconic scene in *North by Northwest* involves a plane swooping at Cary Grant. Despite some people criticizing the scene, Hitchcock said viewers would forgive the illogical assassination attempt due to the excitement of the scene.[9] We agree.

Thornhill disrupts an auction to escape Phillip Vandamm's thugs in *North by Northwest*, and that got us thinking about the human urge of wanting to outbid people. Whether it be in person or online, the thrill of outbidding someone and winning a prize is exhilarating. This is due to the adrenaline rush of competing, the excitement of how quickly things move, and pseudoendowment. This term refers to the feeling of attachment or a sense of ownership for what you're bidding on. People tend to bid higher when they feel pseudoendowment, and it increases their motivation to win at all costs.[10] It's better to go in with a game plan and know exactly how much you can afford to bid so you don't make a rash decision you'll regret. (Kelly says from experience!)

What would the consequences of getting shot with blanks be as seen in *North by Northwest*? Thornhill doesn't die but it's still a dangerous position to be in. A gun loaded with blanks contains gunpowder to create a flash and a bang. If you are any closer than a couple of feet to a gun shooting blanks it could spell danger or even death.[11]

> Mount Rushmore in South Dakota is visited by over three million people per year. About 90 percent of the carving of Mount Rushmore was done using dynamite, and it was completed in 1941.[12]

Eva Saint Marie is an absolute queen scaling a literal monument in a skirt and heels in *North by Northwest*. And people complained about Bryce Dallas Howard running in heels in the *Jurassic World* (2015) movie? Epic! We see Kendall and Thornhill climbing down Mount Rushmore. Is this possible? While it may technically be possible, it is strictly forbidden and considered illegal.

Many other impressive stone carvings are in the world, including sculptures in Nemrut Dağ, Turkey (built between 70 to 38 BCE); the Devil Heads in the Czech Republic, carved between 1841 and 1846; the Bayon Temple in Cambodia, built between 1181 and 1218; and the Crazy Horse Memorial still under construction in South Dakota.[13] Visit Mount Rushmore if you like, but don't expect to complete feats accomplished in this film. You will likely be arrested.

CHAPTER FIFTEEN

Psycho

A midwestern boy named Robert Bloch enters our story. The son of a banker and a social worker, he lived a typical American childhood in Chicago, and later in Milwaukee. He became entranced with horror after seeing Lon Chaney in *The Phantom of the Opera* (1925) reveal his deformed features behind his mask and soon devoted his time to devouring spooky tales in books and on film. We can relate.

Immediately after graduating high school, inspired by H. P. Lovecraft, Bloch began his pursuit of a career in horror, exploring the boundaries of cosmic horror. He was published in numerous magazines and wrote for radio series and TV. As his career burgeoned, Bloch became more interested in writing psychological fare. Before his novel *Psycho*, he had two short stories appear in *Alfred Hitchcock Mystery Magazine* in 1957, in addition to two stories in fiction anthologies under the Hitchcock publishing mantle. These were mere brushes with the director before the explosion of fame that *Psycho* would bring.

As we learned in our research for our first coauthored book *The Science of Monsters* (2019), Robert Bloch, a Wisconsinite himself, was fascinated with the headlines of Wisconsin murderer Ed Gein, who had been captured in 1957. Like so many that winter, he read the newspapers with a grim obsession, captivated by the polite farmhand who had been keeping, in his house, body parts of women he'd murdered and women he'd dug up from their resting places. As we recounted in our first book, this fictionalization of "The Butcher of Plainfield," was a pivotal moment in horror:

Surprisingly, Gein is the spark that ignited both the timid and proper Norman Bates of *Psycho* as well as the mute and brutal Leatherface of *Texas Chainsaw Massacre* (1974). It is the duality of the light and the dark that has placed Norman Bates in the upper echelons of horror film fiends. Robert Bloch's novel, *Psycho*, a novelized account of Ed Gein's house of horrors, was published in 1959. Once Hitchcock read the story of Norman Bates, a man with a mommy complex which charged his sick compulsion to kill women, the famously fastidious director knew he had a subject for his next picture.[1]

If you want to learn more about Ed Gein and read our interview with a Mayo Clinic psychiatrist about the inner workings of Norman Bates's warped psyche, we recommend you pick up a copy of *The Science of Monsters*.

I (Meg) have had a copy of Robert Bloch's *Psycho* on my bookshelf for a while now. It's been waiting patiently for me to pick it up. It's a used copy, dog-eared, with an illustration of a shower drain smeared with blood. Writing this book gave me a good excuse to finally read it, and I can say I was surprised at how similar it is to the film. And Hitchcock's film is in my DNA. Since my mother had me watch it when I was about ten years old, I've probably seen it at least twenty times. I've studied every nuance of Anthony Perkins's face, every quick cut in the shower. So, reading a novel version was surreal.

The plot is essentially the same. Mary (Marion in the movie) steals money from her employer and finds herself at the Bates Motel. She's killed in the shower, although the description is actually more brutal in Bloch's version. In 1960 Hitchcock fought with the MPAA over every single shower shot, as rules were infinitely stricter then than they are now. What the novel was able to provide was more insight into Norman's frame of mind. We get to see behind his mask of politeness a bit more, but I'd say, in this rare instance, the film is better than the book. Screenwriter Joseph Stefano kept Bloch's novel pretty much intact, a rare feat, but it's Hitchcock's attention to building dread and suspense, and Anthony Perkins's and Janet Leigh's acting, that elevates the piece.

Robert Bloch, who passed away in 1994, was a prolific writer who went on to write seventeen episodes of *Alfred Hitchcock Presents*, as well as many more stories and books, but it is *Psycho* that will endure: "In a

November 1985 *Fangoria* interview, Bloch attributed *Psycho*'s long-term impact and ability to frighten to 'the distressing notion that perhaps a boy's best friend may not necessarily be his mother, and the overall message that we may not always know our neighbors as well as we think we do; the most commonplace people in commonplace surroundings can sometimes offer unexpected menace.'"[2]

As experienced moviegoers, we have an innate ability to predict what's going to happen next. If we watch a romantic comedy, there's a good chance the lovers will be split apart before they come together in the end. If it's a slasher movie, there is the final-girl phenomenon in which often the most studious, gentle of female characters is the one left to survive on her own and kill the bad guy with the knife. Alfred Hitchcock knew as he was constructing *Psycho* that by playing with storytelling "rules" he could subvert the audience's expectations, thereby giving us a satisfying and *terrifying* moment when the sudden murder occurs. He discussed this with François Truffaut:

> The audience was probably sorry for the poor girl at the time of her death. In fact, the first part of the story was a red herring. That was deliberate, you see, to detract the viewer's attention in order to heighten the murder. We purposely made the beginning on the long side, with the bit about the theft and her escape, in order to get the audience absorbed with the question of whether she would or would not be caught. Even in the business about the forty thousand dollars was milked to the very end so that the public might wonder what's going to happen to the money. You know the public always likes to be one jump ahead of the story; they like to feel they know what's coming next. So you deliberately play upon this fact to control their thoughts. The more we go into details of the girl's journey, the more the audience becomes absorbed in her flight. That's why so much is made of the motorcycle cop and the change of cars.[3]

In various interviews, Janet Leigh has admitted that because of *Psycho* she developed a lifelong phobia of showers. No joke—she took baths up until her death in 2004! Perhaps it has to do with having to film with her eyes open. Mike Floorwalker wrote about Leigh's experience for *Looper*:

> Hitch's level of preparation and trust in Leigh's actorly instincts meant that the actress was free to give the role her all, resulting in a naturalistic performance quite different from what audiences of the era were used to. Such was her dedication that at the conclusion of the infamous shower sequence in which her character is stabbed to death by Bates, Leigh held her pose—face down on the bathroom floor, eyes wide open as the spray cascades over her—with no help from contact lenses or in-camera trickery.
>
> "[Hitchcock] wanted me to put in those lenses that would give me a scary look. [But it] would've taken six weeks for my eyes to get used to them," Leigh said. "And if I didn't, it could have damaged my eyes. Mr. Hitchcock said, 'You're just going to have to do it on your own.' So I held that look. . . . I will say it wasn't easy."[4]

> You get a gold star, or maybe a Bates stuffed bird, for spotting Alfred Hitchcock in *Psycho*. This is one of his briefest cameos, as he is merely standing outside of Marion's office in a cowboy hat for a few frames as she returns from her steamy lunch break with Sam Loomis (John Gavin).

Last summer, my family and I (Kelly) visited Universal Studios in Hollywood. There were so many incredible Alfred Hitchcock facts and locations on the studio tour ride, and my favorite bit was seeing an actor portraying Norman Bates at a set of the Bates Motel. The house loomed in the distance, and I was grinning ear to ear. It was an incredible experience, and I appreciated all the love I received for my Bates Motel tattoo designed by Josh Kirkpatrick of Duluth Tattoo Company. I made sure to wear shorts that would show it off! Needless to say, *Psycho* had a profound effect on my love of film, and horror in particular.

At Marion's office, the secretary Caroline (Pat Hitchcock) tells the story of how she took tranquilizers on the day of her wedding. What is the history of nervous disorders? Anxiety has always been a part of the human experience, and descriptions of this feeling date back to 5000 BCE.[5] Two thousand years ago, writers began to describe people who suffered from symptoms of anxiety, but it wasn't until the 1600s that doctors gave it a name: *melancholia*. Two hundred years later, the term *panophobia* was used to describe an intense fear that comes at night.[6] Now we would call this night terrors or the Sunday scaries, a feeling of anxiety over returning to work the next morning. By the 1980s specific names were given to anxiety disorders, and they began to be treated.

Marion stops to take a nap on the side of the road while fleeing Phoenix. When she's woken up by a police officer, her demeanor is nervous and flustered. Though she has a legitimate reason to feel this way (she's stolen some money after all), many of us will panic when we see a cop car. Why do we feel anxious around law enforcement? Most of us, when seeing a police car when driving, will automatically hit the brake whether or not we are speeding. This response is programmed into some of us more than others, depending on our past experiences with the law. We may feel fear, respect, or concern when we spot a police car, and it's likely due to how we were brought up and the manner in which those around us spoke about authority.[7]

The car salesman proclaims that the first customer of the day is always the most trouble. Marion puts up no fight and selects a car right away. What is the psychology that usually goes along with purchasing a new vehicle? Most people buy a new car because they need one. Their car may be old or in poor running condition. Others need a different vehicle to fit changing family size, and others buy to get the latest version or technology.[8] The process can be stressful, and salespeople may use some tactics to pressure you, including wearing you down so you don't feel like negotiating, being overly persuasive, and setting a deadline for price or extras. To avoid buying a vehicle you're not able to afford, do your research ahead of time, and stick to your plan. We both admit we should have taken this advice this past year when we needed new vehicles.

Marion places her suitcase on her bed to unpack at the Bates Motel, but recently, people have been saying not to do this. Why is that? The wheels of our suitcases, especially during air travel, touch all kinds of surfaces, including public bathroom floors, sidewalks, and others that we're not familiar with. The second reason is that hotel beds may have bedbugs or other pests that could hitch a ride in your luggage and infest your home.[9] Some estimates guess that anywhere from 10 percent to 20 percent of the hotels in the United States have bedbugs, and one female bedbug can lay up to five hundred eggs in her lifetime.[10] We're convinced to not put our luggage on the bed and are starting this new habit.

Norman, we believe, is taking care of his mother in *Psycho*. He says he doesn't hate her, but he hates what she's become because of her "illness." Marion tries to convince him to put her in an institution, and Norman is deeply offended by the suggestion. What are the statistics for children being caregivers for their parents? According to *Guardian Life*, "one hundred million U.S. adults function as caregivers, providing care for a child, parent, or other relative. More narrowly, there are fifty-three million U.S. adults who care for a spouse, elderly parent or relative, or special-needs child. . . . The number of people in the adult population providing care and working outside the household has increased from one in seven in 2020 to one in five today. And the average time devoted to giving personal care to another has nearly tripled since 2020."[11]

Caregiving can be stressful, but is a necessity for many people who can't afford to hire a personal-care attendant or pay for their loved one to live in an assisted-living facility. While the Family and Medical Leave Act provides certain employees up to twelve weeks of unpaid leave per year to care for a sick family member, most people can't afford to lose this income.[12] As the United States is facing a growing aging population, we hope changes can be made to make care affordable for those who need it.

Norman submerges Marion's car in a body of water to hide the evidence of her murder. There's a moment of anxiety when the car pauses and we, along with Norman, think the car won't be hidden from sight. It sinks, and Anthony Perkins flashes a slight, morbid smile that chills us to the bone. How long would it take a car to sink? Research shows that vehicles will float for thirty seconds to two minutes before completely sinking, so it's important to know what to do if you are in a car that's gone into water. First, try to remain calm then roll down your windows. This will allow you to exit easier. Next, take off your seatbelt before assisting others. Last, climb out a window and swim to safety.[13] We hope that we won't ever have to use these tips!

The habitual eating of candy corn by Anthony Perkins was an acting choice he made for his character in *Psycho*. What's the psychology behind an oral fixation? People who have this obsession constantly need to be sucking or chewing on something due to the focus they have on their mouth. This is an outlet to release nervous energy and self-soothe in stressful situations. An oral fixation can include eating, talking, smoking, drinking alcohol excessively, thumb-sucking, and nail-biting.[14] These habits may cause harm to your teeth or your body, so it's important to seek help to resolve the behavior.

When Detective Arbogast begins to inquire about Marion, Norman starts to stutter. Why do people stutter? Researchers believe stuttering is caused by the following:

A combination of factors, including genetics, language development, environment, as well as brain structure and function. . . . Between 5% and 8% of all preschool aged children will develop stuttering; however, 80% of these children will stop stuttering during early childhood. Certain clinical characteristics have been associated with persistence and recovery of stuttering. Specifically, research suggests that children who are male, have a family history of stuttering, [and] are older at stuttering onset exhibit a higher frequency of stuttering-like disfluencies, produce speech-sound errors, and demonstrate lower expressive and receptive language skills are more likely to continue stuttering.[15]

Norman seemed to be stuttering due to his nervousness when being questioned. He clearly hadn't rehearsed his lies.

Undoubtedly, the score in *Psycho* must be one of the most memorable and iconic of all time. The sound of the strings in the shower scene alone are enough to give us nightmares. We had the opportunity to interview Emmy-nominated, Los Angeles-based composer Holly Amber Church about her work in the industry and how she goes about scoring a film:

Kelly: Tell our readers about your background and what you do in film and television and how you got into the industry.

Holly Amber Church: I am a composer for film and television, meaning I write the musical underscore and/or themes for various projects. I loved film music from the time I was a kid and wanted to be a film composer from a very early age. I studied piano and composition in school, worked for a few established film composers after graduating, and then honed my skills and made industry connections working on various indie films, which led to more and more projects.

Meg: What was your first experience with Hitchcock? Do you have a memory of seeing one of his films for the first time?

Holly Amber Church: I actually think my first roundabout introduction to Hitchcock was that I had the sheet music for "Funeral March of a Marionette," and I would play it on the piano all the time. I eventually realized it was the opening theme song to *Alfred Hitchcock Presents* and thought that was so cool! I think probably one of my earliest Hitchcock film memories was either *North by Northwest* or *The Birds*. I remember *The Birds* being scary because I was kind of afraid of birds as a kid.

Meg: Me too!

Holly Amber Church: We had two parrots in the family when I was growing up, and one of them would try to attack me if it got loose in the house. I still have nightmares about that bird!

Kelly: As a composer, how do you approach scoring a film or television show? What is your process?

Holly Amber Church: I like to really assess and internalize everything I can about a film or television show before I start writing a single note. What's the story? Who are the main characters, and what is important about their individual story arcs? What is the genre and time period of the project, and should any of that play into the score? Should there be a melodic theme for any of the characters or places or any other element of the project? I like to watch the project several times over and over to let it sink in and inspire me, and then I will typically spend a few days or more coming up with melodic theme ideas and an overall sound palette for the project. Is it an orchestral score? Is it a modern sound or a throwback sound? Should it be a synth-based or hybrid score instead? What is the main instrumentation that would work well with the project or any cool or unusual sounds or instruments that could be featured? All of that gives me a very good base and starting point.

Kelly: That's incredible! So much more goes into it than I realized.

Holly Amber Church: I then will typically do what we call a "spotting session" of the film with the director, where we go through the film together and spot where we think music should be and what it should convey in those scenes. Then I am off and running! I tend to like to write linearly, as I have found that I discover things along the way as I'm writing. Then once I get a first pass done, I like to go back to earlier parts of the film and incorporate any new theme ideas or instrumentation choices I may have discovered later in the film that would make sense in some earlier scenes. I, of course, continually send music to the director along the way for their review and feedback as I go and incorporate any thoughts, notes, or ideas from them as well. Writing for film and television is a truly collaborative effort! Then if we will be recording any of the music with live instruments, that typically happens at the very end, and then everything gets finalized.

Meg: What movie soundtracks in particular do you connect with?

Holly Amber Church: Oh, man—I love so many kinds. I do love the horror and thriller genre, so there's a lot of those soundtracks that I really connect with and enjoy listening to. Speaking of Hitchcock, Bernard Herrmann is one of my very favorite composers to listen to, and I love so many of his scores for Hitchcock films (*Psycho* being my favorite, of course). For more recent horror scores, I really love everything that Marco Beltrami writes! On the other side of that, I do also love a big, sweeping orchestral score, like John Williams's *E.T.* (1982) or James Horner's *The Rocketeer* (1991). They're just so incredible!

Kelly: Can you reveal what you are currently working on or what's coming up next for you?

Holly Amber Church: I had a movie come out called *Miranda's Victim* (2023) that I'm really excited for people to see. It tells the story of how the Miranda rights (You have the right to remain silent . . .) came to be. It's a really crazy story that I never knew until I worked on the film, and it's an important story in so many ways. This is the first time that the victim of that case has allowed her story to be told so we all felt an extra sense of duty to work so hard on this film and to do her story justice. It also has an incredible cast that includes Andy Garcia, Luke Wilson, Donald Sutherland, Emily VanCamp, Ryan Phillippe, [and] Kyle MacLachlan and [is] led by Abigail Breslin, who is just extraordinary in her role! It's not horror, but it does have some true crime/thriller aspects to it for sure! I also just finished the score for a documentary about Donn Beach (a.k.a. Don the Beachcomber) called *The Donn of Tiki* that was just a pure joy to work on! We recorded that score with a local tiki band called The Hilo Hi-Flyers, and they sounded fantastic! That should be coming out soon as well. I hopefully have some more fun horror and thriller projects coming up soon too!

Meg: Finally, what is your favorite Alfred Hitchcock film?

Holly Amber Church: My very favorite Alfred Hitchcock film is *Psycho*! It is just marvelous! I can watch that one over and over, and it is a must for me every October. I also just can't get enough of that score!

Music plays such a vital role in film and television, and we will be watching and listening to movies with a new appreciation after speaking with Holly.

If you'd like to learn more about the science of taxidermy and dissociative identity disorder read our book, *The Science of Monsters*, in which we tackle these subjects in relation to *Psycho* and the character of Norman Bates.

CHAPTER SIXTEEN

The Birds

If you search "horror bird movies" in your Google bar, not much comes up. There's an article from the National Audubon Society daring you to watch horrifically bad B-movies like *Birdemic: Shock and Terror* (2010), voted one of the worst movies of the decade. Or how about *Beaks: The Movie* (1987), in which it's stuffed pigeons we're supposed to believe are really pecking people's eyes out. Or even better, *Kaw* (2006), with not-so-great CGI and cringy twists; it's starring Rod Taylor, the buff lawyer Mitch Brenner from *The Birds*. Heard of any of these gems?

In over a hundred years of cinematic history, no one else but the man himself, Alfred Hitchcock, has come close to making birds scary on the big screen. Birds can certainly be scary; I (Meg) am well known to have a dislike of birds in my personal space. I prefer my cardinals and chickadees outside, thank you. If a bird ever flapped close to my neck, I'd probably collapse into a fetal position. And, yes, like many, this is due in part to *The Birds*. I was a kid who loved gory horror movies, but there was something about the brutality of being killed by a hundred pecking birds that disturbed me in particular—like death by a million paper cuts. The shot of the corpse with the pecked-out eye is one of my most vivid childhood remembrances. This must be why I write horror books. I digress—my point is that the reason *The Birds* is considered a masterpiece and isn't in the VHS wastebasket like so many poorly regarded killer-bird movies is because Hitchcock and screenwriter Evan Hunter gave us meaty characters and that all-too-vital pacing of an eyes-glued-to-the-screen thriller.

As with *Psycho*, it seemed important to Hitchcock that the literature his films were being adapted from was respected. "If Hitchcock read the Daphne du Maurier novella only once, his memory for the arc and details of the story was remarkable. Besides the pattern of birds 'attacking, retreating, and massing to attack again,' in the words of Bill Krohn in *Hitchcock at Work*, and the 'vivid descriptions of bird attacks,' several key incidents in the film are lifted directly from du Maurier. These include the discovery of a neighbor found dead with his eyes pecked out, and the climax, with the family barricaded inside a house as the birds rally their final attack—followed by 'the complete absence of explanation for their catastrophe."[1]

We're no stranger to fowl demons of the silver screen. We wrote about Hitchcock's *The Birds* in our book *The Science of Monsters* (2019), in which we discussed how he brought Daphne du Maurier's short story to feathery life, along with some real-life inspiration:

> Before hiring screenwriter Evan Hunter to adapt and tweak the plot of *The Birds*, Hitchcock did his own research on a real-life bird "attack" in Capitola, California. In 1961, a few weeks after this incident, the director requested news copy from the *Santa Cruz Sentinel* to study. He read of the eerie morning of August 18, 1961, when residents of Capitola, California, a smaller community on the coast of Monterey Bay, woke to a frightening discovery. Droves of sooty shearwaters, a medium sized seabird native to the area, were acting erratically; some crashed into rooftops, windows, and cars. Others flopped, dying in the streets, while more vomited their fish dinners into the grass. It was a short incident, but one that terrified and traumatized those who had witnessed this mass bird hysteria.[2]

We hope we've got you intrigued, wondering what happened to those poor birds in Capitola, California. Guess you'll have to grab a copy of *The Science of Monsters* to find out. (Hint—it's a totally scientific explanation, not demon possession or vampire birds.)

> Birds can carry germs that will make people sick, from minor skin infections to serious illnesses.[3] Be careful, friends.

While Hitchcock and Hunter were creating the story, the director needed to find his next Hitchcock Blonde. He wanted Grace Kelly for the part, but she was already in Monaco, married to the Prince and semiretired from acting. He kept an eye on several young actresses for the role, like Sandra Dee, Carol Lynley, and Pamela Tiffin, who'd had a star turn in Billy Wilder's *One, Two, Three* (1961). Yet, it was a single mother from Minnesota whom he ultimately cast as Melanie Daniels. Tippi Hedren appeared in a milk commercial that caught Hitchcock and Alma's eye while they watched the *Today* show. They tracked her down by her modeling agency, and she was given a contract with Alfred Hitchcock, which she believed was for his TV series. She was then put into a sort of Hitchcock class "for several weeks; as Hedren underwent this crash course, she had no idea that she was being considered as the possible leading lady of *The Birds*. She watched *Rebecca*, *Notorious*, and *To Catch a Thief*; then, at home with the Hitchcocks (Alma was usually in the room, observing but saying little, according to Hedren), she rehearsed the very same scenes Hitchcock had famously directed with Joan Fontaine, Ingrid Bergman, and Grace Kelly."[4]

It wasn't until after an expensive three-day screen test of Hedren and another actor (who was not cast) that *The Birds* was even mentioned to her. And this was when she was presented with a gold-and-pearl pin of three birds, in congratulations for her receiving the role.

Tippi Hedren as Melanie is one of our favorite Hitchcock heroines. We're also fans of Veronica Cartwright, who plays Mitch's little sister, Cathy. She grew up to be such a delightful villain in *The Witches of Eastwick* (1987) and has appeared in countless genre fare like the iconic *Alien* (1979).

It's hard to miss Hitch's cameo in *The Birds*, as he's walking out of the pet store with his two adorable Sealyham Terriers at the beginning of the film. The director fell in love with the breed when the actress Madeleine Carroll brought hers on the set of *The 39 Steps*. He owned several over his life, including his cameo partners in *The Birds*, Geoffrey and Stanley.

At the beginning of *The Birds,* Melanie sees many birds flying in the sky over the city. What can we tell about bird behavior when it comes to predicting storms and weather? According to the *Farmer's Almanac,* the old proverbs about birds are true! Here are a few:

- Hawks flying high means a clear sky. When they fly low, prepare for a blow.
- When domestic geese walk east and fly west, expect cold weather.
- If birds in the autumn grow tame, the winter will be too cold for game.
- Geese fly higher in fair weather than in foul.
- Birds singing in the rain indicates fair weather approaching.
- When the swallow's nest is high, the summer is very dry. When the swallow buildeth low, you can safely reap and sow.[5]

In dense, high-pressure conditions, migrating birds have an easier time flying, while in falling air pressure birds will fly lower. Next time a storm is approaching, try to notice if you hear birds singing or not. They tend to be quiet before a storm and make noise once the weather is improving.

You may not want to have a bird as a feathered companion after watching this movie, but we wondered what animal experts thought about keeping birds in the home. Birds have quite a few needs people need to be aware of before committing to getting one as a pet. First, you should be aware of birds' health needs, including what type of food to feed them, bathing practices, beak and claw maintenance, and possible diseases they could be susceptible to. Next, it's important to understand their behavior and body language so you can monitor any changes. Lastly, understand your breed of bird so you know if they are better off as a single pet or prefer companionship. Knowing this will help you choose their habitat, enrichment activities, and socialization expectations.[6]

In *The Birds,* the love birds tilt with the car as Melanie takes the sharp corners up the coastal highway. Classic Hitchcock comedy moment! What is the force that causes this movement? It can be explained by something called centripetal force. The word comes from Latin: *centrum,* meaning "center," and *petere,* meaning "to seek." It's a force that makes a body follow a curved path and was first described mathematically by Dutch physicist Christiaan Huygens in 1659.[7] It may be affected by the fact that

Melanie drives fast wherever she's going. As my (Kelly) grandpa used to say, "She's got a lead foot!"

How many people like to drive fast? According to Leon James, a professor of psychology interviewed by National Public Radio (NPR), we speed because many people are always trying to do things as fast as they can in our culture.[8] While we may feel like speeding will get us places faster, it's also more dangerous. In the United States, nearly 30 percent of car accidents per year are attributed to speeding.[9]

A seagull swooped down and attacked Melanie while she was making her way across the bay in a boat, and a slew of gulls attack at Kathy's (Veronica Cartwright) birthday party. Are birds known to attack people? Though not to the level shown in this movie, most birds will exhibit aggressive behavior if they feel threatened or are guarding their nests. Surprisingly, we discovered that birds kill between seventy-five and one hundred people per year on average worldwide![10] The bird species most likely to attack humans include emus, ostriches, mute swans, cassowaries, and geese. We suggest staying away from these birds if you see them in the wild.

Melanie and Annie (Suzanne Pleshette) exchange notes on Mitch's mother, Lydia (Jessica Tandy), regarding how she seemed cold toward them. What is the psychology behind overprotective mothers? Although we all will feel naturally protective of our children, some parents take this feeling too far. A 2023 psychological study reported that "recollections of maternal overprotective parenting showed small positive correlations with the following schemas: enmeshment, emotional deprivation, mistrust abuse, dependence incompetence, social isolation and alienation, abandonment and instability, failure to achieve, insufficient self-control and entitlement."[11]

Melanie plans to give a mynah bird to her aunt and mentions that these types of birds can talk. How many birds possess this attribute? Dozens of birds are able to mimic human speech, and scientists debate whether they have cognitive understanding of language. The earliest reference to a bird speaking was in the fifth century BCE. It was a bird named Bittacus who belonged to a Greek historian and physician.[12] Talking birds also play a role in many fictional stories, like Edgar Allan Poe's "The Raven"

(1845) and the television series *Twin Peaks* (1990–1991). We admit we'd be a little too spooked to own a talking bird because we've seen too many horror movies.

The building dread and anticipation as crows continue to land on playground equipment outside of the school is one of our favorite scenes in *The Birds*. Annie is remarkably calm as she instructs the children on what to do to leave the school safely. How do we react to demeanor in a stressful situation? As humans, we will naturally react with panic if someone is sharing that type of energy under stress. Instead, people who show composure can help calm others and lead them through the events. If you need to remain calm, try not to let your emotions show, keep a positive attitude, speak with conviction and confidence, and commit to resolving the issue at hand.[13] Following these tips will not only help others but will also allow you to push through without much anxiety.

Mrs. Bundy (Ethel Griffies), an ornithologist, tells everyone at a diner that over one hundred billion birds exist on the planet. Is this accurate? According to *National Geographic*, research estimates that between 50 billion and 430 billion birds are on Earth.[14] This means if all birds attacked, like they do in this film, we would each need to fight off between six and fifty-three of them. We hope this will never happen. Growing up, we were probably two of the only young girls on the planet who believed someday we would need to board up our houses. Watching *The Birds* and *Night of the Living Dead* (1968) made it seem like a real possibility! Thankfully, we're happy to report, it hasn't come to this—yet.

We had the opportunity to interview Caroline Young, author of *Hitchcock's Heroines* (2018), about the filmmaker's complicated past with women who starred in his pictures:

Meg: Could you speak to the controversy of Hitchcock's relationship with the actresses in his films? There are rumors of mistreatment. Did you find that in your research?

Caroline Young: I found in my research that it was a lot more nuanced than the broad brushstrokes of Hitchcock being misogynistic and mistreating his actresses. Tippi Hedren has shared her experiences, and she's covered extensively the traumatic experience she had making *The Birds* and *Marnie*. Other stars had different experiences. Ingrid Bergman and Grace Kelly stayed in touch with him as friends after filming, with Hitchcock and his wife Alma visiting them when in Europe. Kim Novak spoke positively; so did Eva Marie Saint, amongst others. By all accounts, he was shy and socially awkward, he could be brisk in his directorial style, and at times [he created] an uncomfortable situation to provoke a particular performance. But at the same time, I don't want to make excuses for his mistreatment of Tippi. It's a complex question to answer.

Kelly: "We understand. Even as we have been researching and writing about Hitchcock, we have mixed feelings.

Caroline Young: However, the levelling of Hitchcock as a misogynist is not something I necessarily recognize in his films. I find the female characters to be complex, inspiring, and sympathetic. They often showed a breadth of understanding that the male leads lack—and we often see the story unfold from their point of view. Hitchcock, in the 1930s, spoke of how he always kept in mind the power of the female audience, and the type of heroine he wanted to show on screen was one that she should be able to relate to. There are plenty of examples of misogyny in twentieth-century cinema, but I don't think Hitchcock's works should be described as such.

Kelly: Hitchcock subverts our expectations in *Psycho* with a heroine who goes against type—and then dies part way through! How did Janet Leigh's character change the way we look at heroines of suspense?

Caroline Young: It was an incredibly groundbreaking moment to kill off the star of the film thirty minutes in, leaving audiences shattered at the loss of her, as in her screen time she had created a very nuanced, sympathetic character we're rooting for. This shock was one that would be replicated by future directors, including Wes Craven's *Scream* (1996), with Drew Barrymore killed in the first set piece.

Meg: We remember watching Tippi Hedren in *The Birds* and being inspired by her independence and strength. This goes for many of his female characters, like Grace Kelly in *Rear Window*. How did these women stand out in their time? Did moviegoers love them? And how are they remembered now?

Caroline Young: When the films came out, the actresses were very much marketed for their style, for their blonde hair and their wardrobes. So, they were very much part of the appeal for audiences at the time. They enjoyed the suspense and thrills of a Hitchcock film, the touches of macabre comedy, but there were also the aesthetics that the director became renowned for and the joy of seeing stars like Cary Grant and Grace Kelly on the screen."

Kelly: If you had to choose, who is your favorite Hitchcock heroine and why?

Caroline Young: It's difficult to choose, but perhaps Ingrid Bergman in *Notorious*, as she goes through a full redemptive arc, from destructive and hedonistic to mask her pain, to putting her life on the line as an undercover agent. Also, she plays drunk very well.

Meg: We agree!

We encourage you to read Caroline's book *Hitchcock's Heroines* and learn more about these amazing, complicated, and talented women.

> Ravens, crows, jays, and magpies are often considered the most intelligent birds and animals on the planet.[15]

CHAPTER SEVENTEEN

Marnie

In his biography, *Alfred Hitchcock: A Life in Darkness and Light*, Patrick McGilligan shared what life was like in the early 1960s for the world-famous director:

> Standing in the middle of a square in Copenhagen, the director heard a siren blaring and spotted an ambulance heading straight for him. The ambulance screeched to a halt, and out jumped a man, crying, "Autograph, please!" Hitchcock scrawled his autograph, the man jumped back into the ambulance, and off it went, siren blaring. "I don't know if the autograph was for the patient or the driver," he mused later. Another time, stepping off a plane in Tel Aviv and heading down an escalator, Hitchcock was spotted instantly, and the whole airport seemed to pause and look up at him, showering him with applause as he descended.[1]

Even the Papal Swiss Guards in the Vatican hummed the theme song to *Alfred Hitchcock Presents* when he walked by, and children in the remote reaches of Tahiti recognized the master of suspense as he was vacationing there. Alfred Hitchcock's cultural influence was unlike any auteur at the time and still remains to be unrivaled by any film director to date. He was a public figure, representing his films and Hollywood with as much fame as on-screen talent like Cary Grant.

Although the circumstance didn't involve an ambulance, I (Meg) was lucky enough to receive an autograph from *Marnie* star Sean Connery, who wrote "Best Wishes" on a napkin to me, thanks to my dad who knew he was one of my favorite actors when I was a child. It's a prized possession.

Having been successful in creating a media empire that had lasted for decades, Alfred Hitchcock had the distinct advantage of being able to make the movies he wanted to, *how* he wanted to. When Evan Hunter, who had previously worked with him in writing *The Birds*, disagreed with Hitchcock on how to bring *Marnie* to the big screen, he was replaced by Jay Presson Allen, who was willing to include the allusions to rape, which Hunter felt was indecent, from the 1961 novel. Again, Hitchcock and screenwriter Allen stayed rather close to the source material, the novel of a woman plagued with trauma who uses her talents to steal and lie, written by Winston Graham, though the film version of *Marnie* tries to end on a slightly more optimistic note. This is arguable, as Marnie and Mark together are a toxic combination. Perhaps in 1964 it seemed like an appropriate pairing, but with a postfeminist lens it is abhorrent.

When looking back on Hitchcock's oeuvre, *Marnie* isn't one of the more popular films that fans immediately cling to as a favorite. It's divisive and controversial, especially in a more enlightened age; while other films of his seemed to have aged like fine wine (a great example is the box-office disappointment *Vertigo*) *Marnie* is often languishing, though it has a band of loyal supporters.

A 1964 review from the *New York Times* gives us an idea of how the film was perceived when it hit theaters:

This Hitchcockian relationship, explored in sumptuous color, is reminiscent of such memorably maladjusted lovers as Cary Grant and Ingrid Bergman in *Notorious* or James Stewart and Grace Kelly in *Rear Window*. And there's the rub. Hitchcock has taken a pair of attractive and promising young players, Miss Hedren and Mr. Connery, and forced them into roles that cry for the talents of Grace Kelly and Cary Grant. Both work commendably and well—but their inexperience shows. Why, one wonders, did the most reliable of the "big star" directors—a man whose least consequential stories have always had the benefit of the most illustrious players—choose relative newcomers for such demanding assignments? Economy, perhaps? If so, Mr. Hitchcock must plead guilty to pound foolishness, for *Marnie* is a clear miss.[2]

Ouch. The review goes on to complain about other facets of the film, but we'll spare you. It's difficult in this modern age to imagine anyone accusing Sean Connery of being an inexperienced actor, but I suppose we were all young once. Whether you agree with the reviewer's assessment or not, it's true that both Marnie (Hedren) and her husband Mark (Connery) are extremely complicated people with neuroses Sigmund Freud would love to poke at. When Marnie tells Mark she has an aversion to sex, he promises to never touch her—only to then turn around and rape her on their honeymoon. Thus, Marnie tries to kill herself in the swimming pool, and then Mark jumps in and rescues her before she dies. This usurpation of hero by a villain, this complex depiction into psychology, was heavy not just for audiences of the 1960s but also for audiences today.

Don't let all the negativity get you discouraged. A lot of love for *Marnie* is out there. It's considered another Hitchcock marvel of technicolor beauty, and modern reviewers are happy to share their affection, including Richard Brody at the *New Yorker*, who considers it his favorite Hitchcock film: "*Marnie* isn't a horror movie, but it's a movie of horrors, and those horrors are all connected to sex. If there's one constant in Hitchcock's career, it's sex—sexual desire, sexual aversion, sexual fear, sexual repression, sexual gratification—as the engine of human society at both its best (its occasional acts of heroism) and its worst (the crimes that he films with such cunning and such unnerving relish)."[3]

This is where we as women and writers need to report truthfully about accusations leveled at Alfred Hitchcock. His legacy is far from perfect. Tippi Hedren alleges in *Tippi: A Memoir* (2016) that Hitchcock sexually assaulted her on the set of *Marnie* (as well as *The Birds*). This included what she described as unwanted touching, kissing, and manipulative behavior. In his *New Yorker* article Brody mentions that while *Marnie* is still his most beloved Hitchcock work, Hedren's memoir has brought new light to not only the history of the film's making but the context of the film's material as well.

In many ways, *Marnie* was the end of an era. It was the last film with Hitchcock's loyal cinematographer Robert Burks, who passed away in 1968. It is also the last time we hear the iconic collaboration of Hitchcock and composer Bernard Herrmann, who scored the likes of memorable films like *Psycho*.

> We love a spike of the lens (to look directly into the camera)! Hitch gives the camera a glance in the hotel hallway about five minutes into *Marnie*.

Marnie seems disturbed by the red gladiolas in the film. Are people bothered by certain colors or flowers? Chromophobia, an intense fear of colors, often comes with post-traumatic stress disorder. (Spoiler alert—this is the case for Marnie!) People with chromophobia may experience chills, dizziness, excessive sweating, heart palpitations, and trembling when seeing the color(s) they are fearful of.[4] Treatment includes talk therapy, exposure therapy, and medication to help manage panic attacks. Similarly, anthophobia is a fear of flowers and often stems (no pun intended) from a negative experience with flowers, like seeing them at a funeral or getting stung by a bee.

Marnie and her mother agree they don't have a need for any man. How many people today are choosing not to be in relationships or not to marry? According to therapists, there are science-backed reasons to remain single. These include time and space to think, more time for friends and physical activity, opportunities at work, financial freedom, and independence.[5] According to a survey written about in *Forbes* in 2023, 57 percent of single adults in the United States are not looking to date anyone.[6] *Forbes* cited various reasons for the Pew Research Center survey results, but changing societal expectations and gender roles most likely play a role.

> Marnie gets a new job and has keen observation skills. How can we improve our own such skills? "You can develop your powers of observation by going somewhere new or trying something different. This naturally heightens your awareness and focuses your mind. And the more regularly you do this, the more developed your observational skills will become."[7]

Mr. Ward (S. John Launer) has a hard time remembering the five-number safe combination. How many numbers can most people remember? The average person can hold a sequence of seven numbers in their memory. (That is interesting, since phone numbers and social security numbers are both longer.) But tools can help us remember sequences. One trick is to assign numbers a letter and then create a mnemonic phrase to help remember the order:

> Each number is assigned a consonant based on some kind of recognizable relationship between the two:
> 0 = Z or S (zero starts with Z)
> 1 = T or D (one downstroke)
> 2 = N (capital N rotated 90°clockwise resembles the digit 2)
> 3 = M (capital M rotated 90° clockwise resembles the digit 3)
> 4 = R (capital R looks like 4 backward; R is also the last letter of *FOUR*)
> 5 = L (Roman numeral for 50 is L)
> 6 = G (the digit 6 looks like a G)
> 7 = K or C (capital K contains two mirrored 7s)
> 8 = F or B (cursive F and capital B look like a figure 8)
> 9 = P (capital P is a mirror image of the digit 9)[8]

This could be used to remember phone numbers, lock combinations, or other number sequences. Or you could create a clever code with a friend. Perhaps Mr. Ward could have used this technique to remember the number order to the safe. That wouldn't have been good for plot, though, as Marnie got to witness him checking the sequence several times.

Mark Rutland (Sean Connery) studies instinctual behavior and has written about animals as predators. "Lady animals figure very largely as predators," Mark says, and he calls them the "criminal class." Is it true that female animals hunt more than males? Some do. Female lions do most of the hunting in the animal kingdom and cooperate with other females to defend their territory and their cubs. Female spotted hyenas are bigger than their male counterparts and lead battles when conflict arises.[9] And how could we forget the female praying mantis who consumes the male after the deed is done? Slay—literally.

Rutland brings Marnie home to meet his father. They seem to be moving rather quickly! What are the typical stages of relationships? Usually, we begin with the contact stage. This is when you meet someone for the first time and get acquainted. Next is the involvement stage, in which you spend more time with the person, get to know him or her better, and take part in self-disclosure. If you continue to like each other, you'll move to the intimacy stage, where you commit to the relationship, are honest and open with one another, and are public with your status. Every relationship will go through some hardships or disagreements, and this is considered the deterioration stage. If you work on your issues, it's considered repair. If you break up, it's considered dissolution.

Marnie gets caught after stealing money and tells Mark her life story, but he already knows the truth. Why do compulsive thieves steal? They may suffer from kleptomania, which is a rare impulse-control disorder. Kleptomaniacs have the inability to stop taking things that they don't need. Episodes happen without planning and come from a powerful urge to steal. Treatment for this condition includes psychotherapy and medication.[10]

Marnie claims she had never liked men until she met Mark, and some critics of the film believe this an exploration of asexuality. Individuals who identify as asexual don't experience sexual attraction toward anyone from any gender. Asexuality does not equate to celibacy; people who are asexual may choose to engage in sexual acts for reasons other than sexual attraction.[11] Viewers often attribute Marnie's aversion to touch to this. Why do some people feel this aversion? We all have personal boundaries when it comes to distance from others. In general, intimate distance is between zero and eighteen inches. These are the people you feel comfortable being very close to and aren't bothered by them touching you. Personal distance, from eighteen inches to four feet, is reserved for those you are close to but prefer not to touch. Social distance allows more space, four to twelve feet, and is seen in social situations, workplaces, and educational settings. Last, public distance is anywhere from twelve to twenty-five feet, and this is the space that allows you to be farthest away but still hear the person, like in a presentation. Tactile communication, meaning touch, is considered the most primitive of our senses, and those who are uncomfortable with touch avoid it. Touch avoidance could be for a number of reasons, including shyness, fear of germs, being uncomfortable with the person, or personal preference.

Marnie seems to disassociate when going through a traumatic experience. Why do our minds do this? Although disassociation sounds quite intense, it is, in fact, relatively normal. Orygen, the National Centre of Excellence in Youth Mental Health states the following:

> Dissociation is when there is a disruption in the usual way we piece together and connect to the different parts of our world. It can refer to a broad range of complex experiences, so it can be helpful to think of dissociation as on a spectrum or continuum. At one end there is a dissociative experience, such as daydreaming, that everyone experiences, which is common and healthy. At the other end is a highly intense, and often distressing, dissociative experience, which may involve completely disconnecting with reality or even one's own identity.[12]

Those who experience dissociative episodes should seek out a mental-health professional who can help them learn techniques to cope.

Lil Mainwaring (Diane Baker) and Mark wake Marnie up when she's having a nightmare. Should you do this? Experts say no. Because our dreams or nightmares take place during REM (rapid eye movement) sleep, we are unlikely to remember them. Waking someone during this stage will actually ensure that they remember the nightmare. It's better to let them sleep and wake up naturally.[13]

Free association is a technique Marnie thinks Mark wants to use to psychoanalyze her, which leads her to break down crying. How is this used in therapy? Free association could mean someone says a word and then you respond with the first thing that pops into your mind. Or it could be more of a stream-of-consciousness technique in which you begin without a plan or questions and talk freely about the things on your mind. This technique was developed by Sigmund Freud in the nineteenth century and is still used today by some therapists.[14] We can both attest that it is also used as a creative-writing exercise! Perform at your own discretion.

Marnie rides in the fox hunt in the movie, while Mark tries to talk Mr. Strutt out of pressing charges against her for stealing. Marnie ends up shooting her horse after it suffers an injury. Why are horses so difficult to treat? There's a sad but true statistic about racehorses and how many die in a season. The *Los Angeles Times* reported that in a fifteen-week period in 2011–2012, twenty-one horses died in New York. In 2019 thirty horses died at Santa Anita Park, and in 2023 twelve horses died at Churchill Downs.[15] The anatomy of a horse explains why horses are nearly impossible to treat. A thousand-pound body balancing on top of three thin legs (if one is injured) puts pressure on the others, and horses aren't prone to bed rest. They will keep moving and running instinctually. Even if you did get them to rest then they may suffer from colic because of how their intestinal tracts are situated. Open fractures cause even more complications with the possibility of infection.[16] It's sad, but it's a reality in horse racing.

Repressed memories are a theme in *Marnie* due to childhood trauma. How does this affect people later in life? According to the *National Library of Medicine*, "Repressed memory occurs when trauma is too severe to be kept in conscious memory and is removed by repression or dissociation or both. At some later time, it may be recalled, often under innocuous circumstances, and reappears in conscious memory. False memory occurs when a vulnerable patient with a history of over compliant or highly suggestible behavior is unwittingly coached by a respected authority figure to create, as if in memory, an experience that never actually occurred."[17] This certainly explains Marnie's actions and reactions throughout the film.

The film may not be the most popular of Hitch's repertoire; nevertheless, it was fascinating to watch and review his work at this point in his career.

CHAPTER EIGHTEEN

Torn Curtain

Torn Curtain is Alfred Hitchcock's fiftieth film, a feat in the fickle world of Hollywood. It was inspired by British double agent Donald McLean and his wife Melinda Marling, who defected to the Soviet Union with their three children in the early 1950s, after McLean held office in the British Embassy. He was part of a group of spies known as the Cambridge Five. Unlike the hero of *Torn Curtain*, the real-life McLean's allegiance was with the Soviets, as he worked with them as an expert in British policy up until his death in 1983. Hitchcock was fascinated by him living in Moscow with his wife and kids.

The writing of the script went through many hands. It first was offered to Vladimir Nabokov, who had recently adapted his own novel *Lolita* (1955) for the big screen successfully. But Nabokov didn't feel comfortable writing a political thriller. Next, Canadian novelist Brian Moore took a stab at *Torn Curtain*, though apparently Universal and Hitchcock weren't impressed. Though Moore is credited as the writer, it took substantial edits from British writers Willis Hall and Keith Waterhouse to get the screenplay on its feet. Even so, the writing of *Torn Curtain* has been criticized as one of its many missteps.

As a fan of Julie Andrews in her musicals and Paul Newman in *The Outrage* (1964) and *The Prize* (1963), which was similar in theme to *Torn Curtain*, Hitchcock hoped that the casting of his new picture would help its box office, especially with the younger audiences who were now less interested in classic actors like Cary Grant. (Grant appeared in his last film that same year, *Walk, Don't Run*, before retiring from acting.) Unfortunately, as Hitchcock friend and biographer John Russell Taylor recalled, the first meeting between Newman and the director did not go well:

As for Paul Newman, to Hitchcock he was a strange type—and one who got off on the wrong foot with the director. "Hitch invited Newman to a small dinner party," wrote Taylor in his authorized biography. "The first thing Newman did was to take off his jacket at the table and drape it over the back of the chair. Then he refused Hitch's carefully chosen vintage wine and asked for beer instead. And to make matters worse, he insisted on going and getting it himself out of the refrigerator in the kitchen and drinking it from the can."[1]

Many of us would probably find this charming of Newman, a sort of casual, good ol' boy persona that he has cultivated on and off screen. Apparently, Hitchcock did not. And this annoyance only continued as it is widely known that the director and Paul Newman were not friendly on the set of *Torn Curtain*.

Set during the Cold War, *Torn Curtain* stars Newman as an American physicist who pretends to defect to East Germany to gain access to confidential research. Hitchcock was wary about the casting of Newman and Julie Andrews. They weren't cheap, both asking for paychecks of $750,000 for what was a modestly budgeted film. Yet, Lew Wasserman insisted on the pair. Because much was made of the "inexperience" of Tippi Hedren and Sean Connery in *Marnie*, Wasserman made Hitchcock cast the hottest stars at the time. Julie Andrews had become America's sweetheart, thanks to the back-to-back hits of *Mary Poppins* (1964) and *The Sound of Music* (1965), and Paul Newman was highly sought after in a film landscape dominated by masculine action stars, thanks to the popularity of the James Bond franchise. It's true that audiences loved Andrews, Newman, and Hitchcock, but even people on set worried that no chemistry was among any of the key players. There were complaints that Hitchcock had lost his heart and creative inspiration, and this was due in part because the studios had forced his hand in casting. According to *Far Out Magazine*:

Hitchcock also didn't have much patience for the interference of actors in his method. However, that's exactly what happened when Paul Newman joined Hitchcock's 1966 project, *Torn Curtain*. Newman's method acting process might have helped him in other productions, but it ended up angering Hitchcock. The director didn't have time for Newman's constant questioning about the characters, their personal motivations, and the plot, and it led to the two clashing on set. On one occasion, Newman questioned Hitchcock about the motivation of his character, and the director quipped that the "motivation is your salary."[2]

Another reason Hitchcock may have been struggling with the making of *Torn Curtain* was the changing times. Just as the actors considered bankable were different, so were musical tastes. Hitchcock and Bernard Herrmann had made beautiful music together, but with box-office failures like *Marnie* haunting him, and with changing tastes, the director was questioning if Herrmann could provide the modern music *Torn Curtain* needed. There was growing tension between the two. When Hitchcock instructed Herrmann to leave a climactic scene silent, Herrmann didn't listen, conducting music he thought appropriate. A fight ensued, as described on a site dedicated to Bernard Herrmann:

In a decision unheard of in Hollywood, evidently calculated to scold and humiliate Herrmann in front of his peers, Hitchcock dismissed the orchestra midway through recording and cancelled the remaining sessions. After a few terse, embittered words Hitchcock returned to his office, apologized to his employers, fired Herrmann, and offered to pay off the composer out of his own pocket. Hitchcock then telephoned Herrmann, recalls Alan Robinson, a horn player at the recording session, and began berating the stunned composer for stabbing him in the back. No less volatile, Herrmann screamed back that Hitchcock had abandoned his integrity and sold out for an extra couple of bucks. The telephone conversation was brief but deadly, effectively ending one of the most successful artistic relationships in the history of motion pictures. Alfred Hitchcock and Bernard Herrmann never spoke cordially to one another again.[3]

From the beginning, *Torn Curtain* seemed to be a shaky enterprise for Hitchcock. His friends and colleagues described him as feeling uncertain when he was normally so confident, due in large part to the disappointment of *Marnie*. Like all titans of industry, the director was having to contend with the passage of time. The way films were made, the relationships between directors and studio heads, what audiences wanted—everything seemed to be changing. At the end of directing his fiftieth film, Alfred Hitchcock had to start thinking of his legacy.

> Another cute cameo! If you enjoyed the canine companions in *The Birds* then you might just squeal at the adorable baby propped up on Hitch's knee in the lobby of the Hotel d'Angleterre. Also, did you hear? That's the *Alfred Hitchcock Presents* theme song playing!

Paul Newman and Julie Andrews are under bed covers together on the cold ship in the beginning of *Torn Curtain*. From Mulder telling Scully to warm up by climbing into a sleeping bag with him on *The X-Files* (1993–2018) to Bella getting heated up by Jacob in *The Twilight Saga: Eclipse* (2010), we love a good body-heat moment in a love story. Does climbing under the covers with someone really keep us warmer? Yes! But there are other ways to warm up if your body temperature is low. A few quick jumping jacks or walking at a medium pace will increase blood flow without overexerting yourself. Putting your cold hands under your armpits will warm them up, and layering clothing will give your body temperature a boost.[4] That's not as exciting as the sleeping-bag technique, but we'll take it.

> The Iron Curtain is mentioned in the film and refers to the boundary in Europe that separated countries. It began in 1945 after World War II and ended after the Cold War in 1991.

The symbol of pi is used as code in *Torn Curtain*. What does it represent in math? Pi is defined as "the ratio of the circumference of a circle and divided by the distance across, which is its diameter."[5] While the number has been called pi since 1706, it wasn't until 1988 that we started celebrating Pi Day on March 14. A physicist named Larry Shaw started the tradition of eating pie on 3/14 and it's been a hit ever since. Don't make the mistake we have in the past and try to pick up a pie casually on that day. Pies will be sold out! Instead, place your order ahead of time.

The extra-long fight sequence between Hermann Gromek (Wolfgang Kieling) and Michael Armstrong (Paul Newman) with no soundtrack was very powerful and seemed even more realistic with only natural sound. Eventually, Gromek passes out from being held in the gas stove. How long does it take to be affected by gas? It depends on how much gas is in the air and the area of the space one is confined in. "At 25 to 30% gas in air, the oxygen deficiency can cause ringing ears, euphoria, and unexplained behavioral changes. At 50% gas-air mixture, a person taking in a few breaths will be incapacitated and unable to self-rescue. At 75% gas, a person is immediately incapacitated, and death will occur in a matter of minutes."[6]

> *Torn Curtain* features many scenes of characters communicating without shared languages. If you find yourself in this situation, use visual aids, universal gestures, and translation apps to guide your communication. Ultimately, it's best to learn some basic phrases in other languages if you plan to travel.

With the popularity of the 2023 movie *Oppenheimer,* we became interested in the Gamma Five research and science in *Torn Curtain*. In the film, scientists are working on antimissile systems. How does it compare to views today regarding the ethics of nuclear weapons? Joseph Nye, author of *Nuclear Ethics,* addressed the moral and philosophical issues of nuclear weapons. He believed instead of eliminating them we must lower the risk of their use.[7] He continued in this 2023 interview:

When many of us assess morality, we care about three things: the motives and intentions of the actors; the means they use; and the consequences they produce. . . . My four principles for judging moral integrity include: 1. Clarity, logic and consistency; 2. Procedures for protecting impartiality; 3. Initial presumptions in favor of rules and rights; 4. Prudence in calculating consequences. . . . Ethical theory cannot be rounded off and made complete and tidy. As [political theorist and philosopher] Isaiah Berlin once wrote, "Since the ends of men are many, and not all of them are in principle compatible with each other, then the possibility of conflict—and tragedy—can never wholly be eliminated from human life, either personal or social." That is the human condition, but it does not exempt us from making difficult moral choices as we formulate policy.[8]

The danger of nuclear weapons is complicated and terrifying, and the timing of this movie, coming out during the Cold War, most likely made people question their own views.

Mathematics is used quite a bit in *Torn Curtain*, and the plot clearly shows the universality of it. Astronomer and physicist Galileo Galilei said, "[The universe] cannot be read until we have learnt the language and become familiar with the characters in which it is written. It is written in mathematical language, and the letters are triangles, circles and other geometrical figures, without which means it is humanly impossible to comprehend a single word."[9] Poetic! Not everyone agrees that math is a language, but it's proven to bring people to a mutual understanding in this film.

The small, humorous moments, like the older woman being forced on to the bus to keep the pace of the escape, are perfect examples of levity in a moment of high stress in Hitchcock's films. It is hard to believe we would laugh out loud during a movie with such a heavy premise, but that shows Hitchcock's brilliance!

To elude capture in the theater near the end of *Torn Curtain*, Armstrong shouts "Fire!" and causes mass pandemonium. Historically, panics caused by yelling "Fire!" in crowded places led to the deaths of twenty-six people in 1911 and seventy-three people in 1913.[10] Laws were enacted to make the practice illegal, but due to the First Amendment, the constitutional guarantee of free speech, charging someone for a crime or the deaths of victims proved difficult.

Armstrong and Sarah Sherman (Julie Andrews) escape and swim to safety at the end of *Torn Curtain* to presumably live happily ever after—although life after this adventure would surely feel commonplace.

CHAPTER NINETEEN
Family Plot

Like all great stories, this one must come to an end. Alfred Hitchcock's last film arguably brought all his iconic filmic traits together—suspense, dark comedy, and, of course, murder. Even the title *Family Plot* is a double entendre playing with the notion of a cemetery plot and the meaning of plot as plan. Screenwriter Ernest Lehman wanted the film to be on the darker side, yet Hitchcock pushed for it to be cheeky and light. He wanted to end his directing career with a little tongue-in-cheek, which seems appropriate. Hitchcock may have been known to be an exacting, precise auteur, but he was also the delightful host of his TV series who had made a generation of moviegoers point out his cameos, no matter how serious the theme of his films. There were no grand movie stars in *Family Plot*, as Cary Grant and Grace Kelly and many of their contemporaries were retired—and if you read the chapter on *Torn Curtain*, you know Paul Newman wasn't going to be invited back to work with Hitch. Instead, *Family Plot* featured competent genre actors like Karen Black, known for disaster flick *Airport 1975* (1974), horror TV anthology *Trilogy of Terror* (1975), and Robert Altman's *Nashville* (1975). There was also the darkly funny Bruce Dern, who at that time was known for westerns like *Posse* (1975), as well as science-fiction fare like *Silent Running* (1972) and our personal favorite title, *The Incredible Two-Headed Transplant* (1971). We also love him in *The 'Burbs* (1989).

Based on the 1972 novel *The Rainbird Pattern,* by Victor Canning, *Family Plot* centers on conniving family members, one of whom is a fake psychic, who are all trying to kill each other over jewels—you know, the kind of scoundrels who have populated Hitchcock's films for decades!

As a kid, I (Meg) remember watching *Frenzy* (1972), Alfred Hitchcock's second-to-last film. It was about a British serial killer with similarities to Jack the Ripper. At the time I was surprised by the violence depicted, as Hitchcock was not able to be as explicit in his earlier works. This obvious shift is echoed by Eric B. Jones in his article for Medium, as it also relates to *Family Plot*:

> *Family Plot*, as a final film for a career as long as Alfred Hitchcock's, is almost quiet in a way. His prior film, 1972's *Frenzy*, was a return to the thriller and felt very much like a farewell film. *Family Plot* would be released four years later making these films released in the final eight and four years of his life. During my last watch of *Frenzy*, it really came to light how that film felt like Hitchcock finally feeling free to express all the things he presented in his work from years past, but could not due to censors and studio heads. It contains sex, nudity, and sexual violence and turns the formula of the director having his main character being the wrong man in the other direction by having his lead be quite unlikable. In a way, he continued this with *Family Plot.* The first thing about *Family Plot* that you notice is how at ease it is with itself. Hitchcock was dealing with major health issues for both himself and his wife Alma Reville. You notice how the camera doesn't move the way you are accustomed to in a film by him. It feels shot in a manner that is in line with a film from 1976. The score that you would typically hear in a Hitchcock film is by John Williams and it is interesting to hear what is clearly Williams and what is clearly Hitchcock in a fascinating combination of textures.[1]

After his professional breakup with Bernard Herrmann, Hitchcock worked with the legendary Williams, as Jones noted, who had just come off his memorable turn on *Jaws* (1975). Music was such a vital piece of Hitchcock's films, as music and suspense go hand in hand, that Williams makes a new kind of mark on the director's last film. Jones also noted that Alma was in struggling health while *Family Plot* was shot. She had been suffering from breast cancer, though she survived it.

After the release of *Family Plot,* Alfred Hitchcock lived for four more years. In 1979 he was the seventh person honored with the American

Film Institute's Life Achievement Award. In his speech, he gave tribute to Alma, with whom he had shared both his personal and professional life for over fifty years. That same year he was named Knight Commander of the Order of the British Empire by Queen Elizabeth II. He died April 29, 1980. His beloved Alma died two years after her husband on July 6, 1982.

We leave you with an excerpt of Hitch's obituary published in the *Washington Post*:

Sir Alfred Hitchock, eighty, the British-born director who for fifty years frightened and delighted movie audiences with thrillers that set screen standards for terror and suspense, died Tuesday morning at his home in Los Angeles. Although the cause of death was not immediately announced, Mr. Hitchcock had a heart pacemaker and had been suffering from both kidney failure and arthritis. No movie director in the history of film was more popular with audiences, more consistently successful at the box office nor more prankishly public a figure than Alfred Joseph Hitchcock, whose name was more prominently displayed than those of the stars on most of the films he made. Mr. Hitchcock became a genre unto himself, commonly known as "the master of suspense," and an unchallenged whiz at manipulating movie and television audiences with tales of fear and mystery. Film critic Pauline Kael has said of him, "A pretty good case could be made for Alfred Hitchcock as the master entertainer of the movie medium." Yesterday, some of the actors who appeared in some of Mr. Hitchcock's fifty-three films both mourned and praised him. "There was nobody like him, and he'll be very hard to replace," said James Stewart, the pursued and beleaguered Hitchcock hero of *Rear Window, Vertigo* and *The Man Who Knew Too Much*. "I've lost a wonderful friend," Stewart said. "The world has lost a tremendous contribution to the art of film and to millions and millions of people." Janet Leigh, the actress who played the victim of the most famous murder in screen history—the shower stabbing in *Psycho*—called Mr. Hitchcock "a master of his profession," and Anthony Perkins, who played the psycho with the knife, recalled of Mr. Hitchcock that "he always executed his suspense with taste; he never offended you. You were scared by it, but pleasantly."[2]

> One last cameo by Hitchcock is seen about forty minutes into *Family Plot*. That instantly recognizable profile in shadow is seen behind the glass door marked REGISTRAR OF BIRTHS AND DEATHS.

Blanche works as a psychic in *Family Plot*. What is the history of fortune tellers throughout the world? People have been predicting the future for centuries using a variety of methods, like tarot card reading, numerology, astrology, palm reading, and clairvoyance. Fortune telling and other psychic services are a $2 billion industry in the United States, and it continues to grow. According to one study, over a quarter of people believe that humans have psychic abilities.[3] As revealed in the film, not all psychics are real. How do fake psychics trick people into believing them? Some people may use vague and general information that allows people to relate to it somehow and believe it's about them. Oftentimes, the specifics are impossible to validate, and those who visit psychics are already biased into believing what the professional says is true.[4]

The trader (Karen Black) doesn't speak at all when making the exchange for the jewel in *Family Plot*. How recognizable are people's voices? We all have speech patterns that are specific to us, including pitch (how high or low your voice is), rate (how fast or slow you talk), volume (how loud or soft you speak), and quality (what your voice sounds like). We may change our speech patterns when we're nervous, tired, or in special contexts that require a more relaxed demeanor or a formal one. Babies begin to recognize familiar voices between the age of one to three months and respond more to high pitched voices.[5] So that explains baby talk! And, we admit, we may speak to our pets this way too.

> What do experts look for to price and authenticate
> jewels? The valuation is determined by carat (the
> weight equivalent to one-fifth of a gram), clarity
> (internal landscape and external blemishes), cut
> (shape, finishing touches, and polish), and color
> (hue, saturation, and tone of the stone).[6]

The hostage in *Family Plot* was given ketamine. How does this drug work? Ketamine is a "dissociative anesthetic that has some hallucinogenic effects. It distorts perceptions of sight and sound and makes the user feel disconnected and not in control."[7] The drug has been used in the past several decades to treat depression but has also been abused as a street drug. The effects of the drug last between thirty and sixty minutes, and this tracks in its use in the film to temporarily impair the hostages.

Cleverly, the jewel is hiding in plain sight in *Family Plot*. How does the human mind work when trying to spot things? The notion of hiding out in the open began in the 1600s on the battlefield when soldiers positioned themselves and used camouflage. Edgar Allen Poe wrote, "The best place to hide something is often right out in the open."[8] Much of this is due to human perception. If your families are anything like ours, your spouse or children can be looking for something and be unable to locate it. When you come to help, it's right in front of their nose. Why is this? We naturally filter information through our senses so as not to be overwhelmed. When we are overly focused on one thing, we may unintentionally ignore the bigger picture. This may explain why our families can't find anything when they're looking for it—but it doesn't excuse it! Just kidding. Someday our kids will have their own homes and understand the struggle.

> To make the real cemetery in the film appear
> unkempt, production paid the property to forego
> lawn maintenance for four months.[9]

Family Plot is a clever title as it can have double meaning. One clear meaning is that of a plot of land in a cemetery. How are burial and funeral statistics changing in the modern era? More people have been choosing cremation over burial since 2015 for a variety of reasons, and the National Funeral Directors Association predicts that by 2036, 80 percent of people will choose it.[10] One of the main driving factors is the cost. A casket, plot, and headstone can be quite expensive, whereas cremation can be a fourth of the cost. Ashes can be kept in an urn, spread in a special location, or even turned into jewelry or fireworks.

The cemetery is practically a character in this film, and we began to wonder how graves have been marked throughout history. In colonial America, graves were marked with wood planks or field stones. It was believed that these items would prevent the dead from rising from the grave. By the seventeenth and eighteenth centuries, engraved headstones were common and became more ornate as time went on. The nineteenth-century grave markers were more personalized and represented the deceased's life. Currently, headstones range from etchings to photos fashioned using computer software.[11] It will be interesting to see how technology will continue to evolve to honor the dead.

The man who sold the headstone for Edward Shoebridge, a.k.a. Arthur Adamson (William Devane), begins to remember the man who ordered it. How can we jog our memories? Beginning at the age of thirty, our memories start to fade, and we become more forgetful. We don't think it will happen to us, but, alas, it does. To retrieve old memories, it's important to engage the senses. What were you eating at the time? What did you smell? Was a song playing? Do you have pictures of the event? Using these techniques, your mind is more likely to remember things. Another technique can be to write things down in a journal and revisit these notes.[12]

Mr. Adamson removes a piece of lint from the detective's suit in a scene. It's a rather intimate gesture considering they just met, as such is usually regarded as a "tie sign" shared between people who know each other well. Tie signs can be an object, like an engagement ring, or a public display of affection. By performing a ritual usually reserved for someone close, Adamson may have been subtly appealing to the detective's subconscious.

The runaway vehicle on the windy road is enough to get our hearts pounding in *Family Plot*, but the comedic elements of the couple reacting lightened the mood. What should you do if you're in a vehicle and can't stop? Don't do as Blanche Tyler did and put your feet in the air. Instead, don't panic—easier said than done. Check if there's anything underneath the brake, like a can or floormat, and remove it. If there's nothing there, try pumping your brakes or pushing the brake pedal all the way to the floor. If that doesn't work, shift into a lower gear to slow down, and engage the emergency brake. Last, carefully create some friction by rubbing against a guard rail and exit the road.[13] I hope none of us will need to use this advice!

By the end of the film, Blanche appears to be in a trance and locates the jewels in the chandelier. What causes trancelike behavior in humans? Someone in a trance is unresponsive and not self-aware. An individual may go into a trancelike state for various reasons, including intense focus, participation in battle, hypnotism, or communication with the "other side." Advancements in science have enabled trance states to be observed through brainwaves; further, one's "knowledge of brainwave states enhances a person's ability to make use of the specialized characteristics of those states: these include being mentally productive across a wide range of activities, such as being intensely focused, relaxed, creative and in restful sleep."[14] The brainwave states include high-amplitude, low-frequency delta to low-amplitude, high-frequency beta. Think of these states ranging from deep, dreamless sleep to a state of high arousal.

What about trancelike states in animals? I (Kelly) grew up hearing about my mom and her cousin hypnotizing chickens by covering their eyes and petting them slowly. What is the science behind this? Evolutionary biologists hypothesize that these trance states could be a defense against predators. Many animals play dead to avoid actual death, including possums, rabbits, ducks, and lemon sharks.[15] If you encounter an animal playing dead, it's best to leave it alone.

Blanche "broke the fourth wall" in *Family Plot* by winking at the camera at the end of the movie. What narrative purpose does this serve in film, television, and theatre? The phrase was coined by French playwright Moliere and refers to acknowledging the audience to bring it into better understanding of a character and the story being told. Contemporary film and television examples of breaking the fourth wall include *Ferris Bueller's Day Off* (1986), *Deadpool* (2016), and *Fleabag* (2016–2019). We feel in on private knowledge through this writing choice and can relate to the character(s) speaking to us.

And now we break the fourth wall with you, dear readers, and hope you enjoyed learning more about Alfred Hitchcock and the science behind his filmography. We wish we could have covered every one of his brilliant works, but there are only so many pages for this book.

Conclusion

As we took on the privilege of writing this book, we wanted to make sure to not only view Hitchcock's films through the lens of the era in which they were created but also see if they live up to our expectations today. We spoke with Madeline Di Nonno, president and CEO of the Geena Davis Institute (GDI), about her experience with Hitchcock and her organization's mission in film and television. Earlier in her career, she had a hand in releasing many of Hitchcock's films and television series, initially on VHS and then DVD. First, the organization needed to get the physical films from an actual basement and protect them. Next, it engaged the film preservation team at UCLA; it was a massive undertaking. Now, we are able to enjoy these important films in our own homes either through physical media or digitally.

Madeline said, "The Geena Davis Institute's work is focused on the power of storytelling to change the world. Authentic representation has great impact on our social and cultural beliefs, our behaviors, and how we see ourselves in the world."[1] She went on to say that Hitchcock always had strong female leads in his films and the women didn't necessarily wait for a man to save them.

We see film and television trending in positive ways in representation and diversity and applaud Hitchcock for the riveting roles he provided for women. We also acknowledge the antiquated, problematic parts of his past and hope he would be more empathetic today. *The Science of Alfred Hitchcock*, written from our modern lens, is about discovering the science and history behind one of the most influential and prolific filmmakers of the twentieth century. Hitchcock's ouvre is still remarkably pertinent to world cinema, and his work will endure.

Acknowledgments

Thank you to all the incredible people who took time out of their busy schedules to be interviewed for this book! We appreciate you and your expertise.

Thank you to our families for their constant support and the never-ending movie nights of rewatching Alfred Hitchcock over the past several months. A special thanks to Campbell for being our resident historian and fellow movie watcher.

We are grateful for everything our team has done to support and promote us. This includes Karmen, Stacey, Mickey, Matt, and Ryan. We appreciate you!

Thanks to the outstanding team at Skyhorse, including Nicole M., Nicole F., Michael, and everyone else who worked so hard to make this book what it is.

And as always, to our Rewinders, we'll see you in the horror section!

About the Authors

Bram Stoker Award–nominated author Meg Hafdahl is the creator of numerous stories and books. Her fiction has appeared in anthologies such as *Eve's Requiem: Tales of Women, Mystery and Horror* and *Eclectically Criminal*. Her work has been produced for audio by The Wicked Library and The Lift, and she is the author of three popular short-story collections, including *Twisted Reveries: Thirteen Tales of the Macabre*. Meg is also the author of the three novels *The Darkest Hunger, Daughters of Darkness*, and *Her Dark Inheritance*, which was called "an intricate tale of betrayal, murder, and small-town intrigue" by Horror Addicts and "every bit as page turning as any King novel" by *RW Magazine*. Meg, also the cohost of the podcast *Horror Rewind* and coauthor of *The Science of Monsters, The Science of Women in Horror, The Science of Stephen King, The Science of Serial Killers, The Science of Witchcraft*, and *The Science of Agatha Christie*, lives in the snowy bluffs of Minnesota.

Kelly Florence is the Bram Stoker–nominated coauthor of seven books in *The Science of . . .* series as well as *Strange and Spooky Spots Across America*. She is the cohost of the *Horror Rewind* podcast and the creator of the *Be a Better Communicator* podcast. She is a communications professor at Lake Superior College in Duluth, Minnesota. Kelly received her BA in theater from the University of Minnesota-Duluth and her MA in communicating arts from the University of Wisconsin-Superior. She is part of the Minnesota Film Critics Alliance and is a programmer for the Minnesota Film Festival. She is passionate about female representation in all media, particularly the horror genre.

Endnotes

Chapter One: The Lodger

1. Patrick McGilligan, *Alfred Hitchcock: A Life in Darkness and Light* (New York: Harper Collins, 2003), 7.
2. "Phobias," Cleveland Clinic, n.d., https://my.clevelandclinic.org/health/diseases/24757-phobias.
3. McGilligan, *Alfred Hitchcock*, 27.
4. McGilligan, *Alfred Hitchcock*, 46.
5. Ivor Montagu, "Working with Hitchcock," *Sight and Sound,* Summer 1980, 189.
6. John Pruitt, "Between Theater and Cinema: Silent Film Accompaniment in the 1920s," American Symphony Orchestra (website), n.d., https://americansymphony.org/concert-notes/between-theater-and-cinema-silent-film-accompaniment-in-the-1920s/. December 19, 1993.
7. David and Hayley A. Young, "Do Small Differences in Hydration Status Affect Mood and Mental Performance?," Supplement, *Nutrition Reviews* 73, no. S2 (September 2015), S283–96, https://doi.org/10.1093/nutrit/nuv045.
8. Leonard G. Johns et al., *Serial Murder: Multi-Disciplinary Perspectives for Investigators*, ed. Robert J. Morton and Mark A. Hilts (Washington, DC: Department of Justice, n.d.), 14, https://www.fbi.gov/file-repository/stats-services-publications-serial-murder-serial-murder-july-2008-pdf.
9. Anjali Sayee "Bleaching Hair with Hydrogen Peroxide at Home—a Tutorial," StyleCraze (website), last modified January 10, 2025, https://www.stylecraze.com/articles/hydrogen-peroxide-hair-bleach/#how-to-bleach-hair-with-hydrogen-peroxide-and-baking-soda.
10. Devon Hopp, "The Extraordinary History of Hair Color," Byrdie (website), last modified September 24, 2024, https://www.byrdie.com/hair-color-history.
11. Jan Hicks, "William Henry Perkin and the World's First Synthetic Dye," *Science and Industry Museum Blog*, August 25, 2017, https://blog.science-andindustrymuseum.org.uk/worlds-first-synthetic-dye/.
12. Hopp, "Extraordinary History of Hair Color."
13. Lucy Jane Miller et al., "Identification of Sensory Processing and Integration Symptom Clusters: A Preliminary Study," *Occupational Therapy International* (November 16, 2017): https://doi.org/10.1155/2017/2876080.

14. J. I. P. de Vris, G. H. A. Visser, and H. F. R. Prechtl, "The Emergence of Fetal Behaviour. I. Qualitative Aspects," *Early Human Development* 7, no. 4 (December 30, 1982): 301–22, https://doi.org/10.1016/0378-3782(82)90033-0.

Chapter Two: Blackmail

1. "How Were Silent Film Intertitles Used?," Beverly Boy Productions (website), December 18, 2024, https://beverlyboy.com/filmmaking/how-were-silent-film-intertitles-used/.
2. "American Silent Feature Film Database," Library of Congress, n.d., https://www.loc.gov/programs/national-film-preservation-board/preservation-research/silent-film-database/.
3. "Becoming a Film Archivist: Essential Steps for a Successful Career," Yellowbrick (website), July 10, 2023, https://www.yellowbrick.co/blog/film/becoming-a-film-archivist-essential-steps-for-a-successful-career.
4. Donald Spoto, "Alfred Hitchcock, Designer," *Print*, July/August 1977, https://www.printmag.com/design-culture/alfred-hitchcock-designer/.
5. Spoto, "Alfred Hitchcock, Designer."
6. Fritzi Kramer, "Blackmail (1929) a Silent Film Review," *Movies Silently* (blog), January 1, 2017, https://moviessilently.com/2017/01/01/blackmail-1929-a-silent-film-review/.
7. McGilligan, *Alfred Hitchcock*, 113.
8. Chris Wiegand, "Blackmail Review—Play That Brought Hitchcock a Hit Is Retooled for Today," *The Guardian*, March 10, 2022, https://www.theguardian.com/stage/2022/mar/10/blackmail-review-hitchcock-mark-ravenhill.
9. Paul McGuire Grimes, discussion with the authors, December 29, 2023.

Chapter Three: *The 39 Steps*

1. McGilligan, *Alfred Hitchcock*, 26.
2. "Here's What to Expect at MEPS," U.S. Army, n.d., https://www.goarmy.com/how-to-join/steps/processing-station.
3. Andre Sennwald, "Alfred Hitchcock's New Picture: The Thirty-Nine Steps," *New York Times*, September 14, 1935.
4. "French New Wave," Criterion Channel, n.d., https://www.criterionchannel.com/french-new-wave.
5. François Truffaut, *Hitchcock/Truffaut* (New York: Simon and Schuster, 1983), 99.
6. "Memory Loss: 7 Tips to Improve Your Memory," Mayo Clinic, March 7, 2024, https://www.mayoclinic.org/healthy-lifestyle/healthy-aging/in-depth/memory-loss/art-20046518.
7. "How to Build a Memory Palace," Art of Memory (website), last modified April 2, 2023, https://artofmemory.com/blog/how-to-build-a-memory-palace/.

8. Surraj Susai, Mrudula Chandrupatla, and Rohini Motwani "Anatomy Acts Concerning Body and Organ Donations Across the Globe: Past, Present, and Future with a Special Emphasis on the Indian Scenario," *Anatomy and Cell Biology* 56, no. 1 (March 31, 2023): 1–8, https://doi.org/10.5115/acb.22.166.

9. "Why Men Fight and What it Says About Masculinity." Atkinson, Scott. September 12, 2017. *The Guardian.*

10. Megan Hoffer et al., "How Music Affects Your Mind, Mood, and Body," Tallahassee Memorial Healthcare, December 2, 2022, https://www.tmh.org/healthy-living/blogs/healthy-living/how-music-affects-your-mind-mood-and-body.

11. "Celebrating Women in Espionage," Spy Museum, n.d., https://www.spymuseum.org/exhibition-experiences/sisterhood-of-spies/

12. Brett Rains, "Oklahoma Police Show How Textbooks Can Stop Bullets," 40 29 News, last modified March 1, 2018, https://www.4029tv.com/article/oklahoma-police-show-how-textbooks-can-stop-bullets/19048296.

13. Marc Griffiths, "'Unleashed, Mad, and Dangerous:' How Britain's Wild, Romantic Moorland Is Our 'Signature Habitat,' Inspiring Everything from Beowulf to Hound of the Baskervilles," *Country Life*, June 13, 2020. Online article, retrieved January 12, 2024.

14. "Is Handedness Determined by Genetics?," MedlinePlus, n.d., https://medlineplus.gov/genetics/understanding/traits/handedness/.

15. "Is Handedness Determined by Genetics?," MedlinePlus.

16. "Do You Really Get Sick From Being Cold?," Unity Point Health, n.d., https://www.unitypoint.org/news-and-articles/do-you-really-get-sick-from-being-cold.

17. "What is Lima Syndrome?" Saladi-Schulmna, Jill Ph.D. Heathline.com. October 20, 2020.

18. Yvaine Ye, "Photographic Memory," *New Scientist*, n.d., https://www.newscientist.com/definition/photographic-memory/.

Chapter Four: *The Lady Vanishes*

1. Alfred Hitchcock, "Alfred Hitchcock and Peter Bogdanovich (1963)," interview by Peter Bogdanovich, The Hitchcock Zone (website), 1963, https://the.hitchcock.zone/wiki/Alfred_Hitchcock_and_Peter_Bogdanovich_(1963).

2. McGilligan, *Alfred Hitchcock*, 208.

3. "How Impossible Worlds Came Alive in Movies—The Story of Miniatures," SrushtiVFX (website), November 20, 2023, (website discontinued).

4. George E. Turner, "Alfred Hitchcock's Rope—Something Different," *American Cinematographer*, January 16, 2023, https://theasc.com/articles/alfred-hitchcock-rope.

5. Jeffrey Michael Bays, "Humor: Hitchcock's Secret Weapon," Medium (website), April 23, 2021, https://medium.com/life-and-the-performing-arts/humor-hitchcocks-secret-weapon-94a0bde4e5e6.

6. "Wind Explained," U.S. Energy Information Administration, n.d., https://www.eia.gov/energyexplained/wind/.

7. David Johnson, "How Tall Would You Have Been 100 Years Ago?," *Time*, July 27, 2016.

8. Diana Deutsch, "The Processing of Pitch Combinations," in *The Psychology of Music*, ed. Diana Deutsch (London: Elsevier, 2013), 249.

9. Tim Tedeschi, "The Scientific Reason Songs Get Stuck in Your Head," University of Cincinnati, June 21, 2024, https://www.uc.edu/news/articles/2024/06/the-scientific-reason-songs-get-stuck-in-your-head.html.

10. Bernhard Fink et al., "Evolution and Functions of Human Dance," *Evolution and Human Behavior* 42, no. 4 (July 2021): 351–60, https://doi.org/10.1016/j.evolhumbehav.2021.01.003.

11. Robert Krulwich, "The List of Animals Who Can Truly, Really Dance Is Very Short. Who's on It?," NPR, April 1, 2014, https://www.npr.org/sections/krulwich/2014/04/01/297686709/the-list-of-animals-who-can-truly-really-dance-is-very-short-who-s-on-it.

Chapter Five: Rebecca

1. Kellaway, Kate. (April 14, 2007) "Daphne's Unruly Passion." *The Guardian*.

2. McGilligan, *Alfred Hitchcock*, 222.

3. Truffaut, *Hitchcock/Truffaut*, 127.

4. Jay Jacobson, "39. Rebecca, 1940," *Jay's Classic Movie Blog*, January 12, 2021, https://www.jaysclassicmovieblog.com/post/45-rebecca-1940.

5. Dennis Perry, "David Hitchcock and Alfred Selznick: Will the Real Director of Rebecca Please Stand Up?" Brigham Young University College of Humanities, 2018. Eric Baker provided his recollection.

6. McGilligan, *Alfred Hitchcock*, 53.

7. Frank S. Nugent, "Splendid Film of du Maurier's *Rebecca* Is Shown at the Music Hall," *New York Times*, March 29, 1940, https://www.nytimes.com/1940/03/29/archives/the-screen-splendid-film-of-du-mauriers-rebecca-is-shown-at-the.html.

8. Eric Suni, "Dreams: Why They Happen & What They Mean," Sleep Foundation (website), May 2, 2024, https://www.sleepfoundation.org/dreams.

9. George Gissing, *The Odd Women* (Peterborough, ON: Broadview Press, 1998).

10. Rebecca Zucker, "How to Succeed When You Have Big Shoes to Fill," *Harvard Business Review*, February 17, 2020, https://hbr.org/2020/02/how-to-succeed-when-you-have-big-shoes-to-fill.

11. Zucker, "How to Succeed."

12. Vilayanur S. Ramachandran and Baland Jalal, "The Evolutionary Psychology of Envy and Jealousy," *Frontiers in Psychology* 8 (September 18, 2017): https://doi.org/10.3389/fpsyg.2017.01619.

Chapter Six: Suspicion

1. Dean Talbot, "Impact of Book Publishing on Film Industry," WordsRated (website), February 23, 2023, https://wordsrated.com/impact-of-book-publishing-on-film-industry/.
2. Alfred Hitchcock, "Suspicion: Proof That a True Artist Can Create Great Things Even under the Crippling Harness of a Profit-Driven Studio," interview by Bryan Forbes, Cinephilia & Beyond (website), October 3, 1969, https://cinephiliabeyond.org/alfred-hitchcock-suspicion/.
3. Truffaut, *Hitchcock/Truffaut*, 142.
4. Jenna Fletcher, "What is Hyperosmia and What Causes It?," Medical News Today, last modified July 18, 2023, https://www.medicalnewstoday.com/articles/321937.
5. Jennifer A. Wade, "(I Think) You Are Pretty: A Behavior Analytic Conceptualization of Flirtation," *Perspectives on Behavioral Science* 41, no. 2 (June 6, 2018): 615–36, https://doi.org/10.1007/s40614-018-0136-y.
6. Jen Kim, "How Animals Flirt and What We Can Learn From Them," *Psychology Today*, November 21, 2021, https://www.psychologytoday.com/us/blog/valley-girl-brain/202111/new-lessons-in-creative-flirting.
7. Kim, "How Animals Flirt."
8. M. M. Moore, "Nonverbal Courtship Patterns in Women: Contact and Consequences," *Ethology and Sociobiology*, 6, no. 4 (1985): 237–47, https://doi.org/10.1016/0162-3095(85)90016-0.
9. "National Survey Finds Most Engaged Couples Are Interested in Eloping This Year," PR Newswire, July 11, 2022, https://www.prnewswire.com/news-releases/national-survey-finds-most-engaged-couples-are-interested-in-eloping-this-year-301583425.html.
10. Anne Fabiny, "Music Can Boost Memory and Mood," Harvard Health Publishing (website), February 14, 2015, https://www.health.harvard.edu/mind-and-mood/music-can-boost-memory-and-mood.
11. Patti Nasier, "Music Therapy for Dementia: How Music Can Benefit Dementia & Alzheimer's Patients," Senior Home Transitions (website), n.d., https://seniorht.com/music-therapy-dementia-alzheimers/.
12. Jordan Greene, "12 Relationship Red Flags You Should Never Ignore, According to Experts," Today.com, November 1, 2022 (page removed).
13. "Compulsive Gambling," Mayo Clinic, July 18, 2022, https://www.mayoclinic.org/diseases-conditions/compulsive-gambling/symptoms-causes/syc-20355178.
14. Gary D. Bond et al., "Lie-Biased Decision Making in Prison," *Communication Reports* 18, no. 1-2 (2005): 9–19, https://doi.org/10.1080/08934210500084180.
15. Trisha Ahmed, "Poison Specialist and Former Medical Resident at Mayo Clinic is Charged with Poisoning His Wife," Associated Press News, last modified October 24, 2023, https://apnews.com/article/mayo-clinic-resident-poison-wife-9b8e8c259316dfbf0c7ea20f4624f0aa.

Chapter Seven: *Shadow of a Doubt*

1. Roger Ebert, "Uncle Charlie Brings Excitement to a Small Town," RogerEbert.com, November 9, 2011, https://www.rogerebert.com/reviews/great-movie-shadow-of-a-doubt-1943.

2. Andre Soares, "Hitchcock Heroine Teresa Wright: Shadow of a Doubt, What Lurks Beneath," AltFilmGuide, 2014, https://www.altfg.com/hitchcock-heroine/.

3. Cara Goodwin, "The Research on Baby Names," *Psychology Today*, July 26, 2023, https://www.psychologytoday.com/us/blog/parenting-translator/202307/the-research-on-baby-names.

4. Edna May Wonacott, "Interview with Edna Green (Edna May Wonacott) from Hitchcock's *Shadow of a Doubt*," interview by Steve Wemisse, *Alfred Hitch-blog*, April 17, 2012, https://alfredhitchblog.wordpress.com/2012/04/17/interview-with-edna-green-edna-may-wonacott-from-hitchcocks-shadow-of-a-doubt/.

5. "Earle Leonard Nelson," *Murderpedia* (blog), n.d., https://murderpedia.org/male.N/n/nelson-earle.htm.

6. National Heart, Lung, and Blood Institute, "Researchers Study How Daytime Naps May Influence Health," National Institutes of Health, May 11, 2023, https://www.nhlbi.nih.gov/news/2023/researchers-study-how-daytime-naps-may-influence-health.

7. National Heart, Lung, and Blood Institute, "Researchers Study."

8. *Encyclopedia Britannica*, s.v. "Cipher," https://www.britannica.com/topic/cipher.

9. "In a Funk? Here's How to Shake It Off," WebMD, https://www.webmd.com/balance/ss/slideshow-feel-better.

10. "How Many Questions Children Ask in a Year . . . a Lot!" 107.5 KOOL FM, November 29, 2023, https://1075koolfm.com/how-many-questions-children-ask-in-a-year/.

11. "Knowing the Facts of Youngest Character, Spoiled Is Not Always True," VOI (website), https://voi.id/en/lifestyle/212149.

12. Felix Planer, *Superstition* (London: Cassell, 1980).

13. "Hat Myths," Hats in the Belfry (website), n.d., https://www.hatsinthebelfry.com/pages/hat-myths?.

14. Catherine L. W. Striley, Roland R. Griffiths, and Linda B. Cottler, "Evaluating Dependence Criteria for Caffeine," *Journal of Caffeine Research* 1, no. 4 (December 2011): 219–25, https://doi.org/10.1089/jcr.2011.0029.

15. Striley, Griffiths, and Cottler, "Evaluating Dependence Criteria."

16. "Baking Science Explained: 3 Chemical Reactions Needed to Bake a Cake," KiwiCo (website), April 16, 2023, https://www.kiwico.com/blog/the-science-behind/baking-science-chemistry-baking-cake?.

17. Dimitrije Curci, "How Many Books Does the Average Person Read?," Wordsrated (website), October 31, 2023, https://wordsrated.com/how-many-books-does-the-average-person-read/.

18. "Top 10 Benefits of Reading for All Ages," Markham Public Library, August 6, 2020, https://markhampubliclibrary.ca/blogs/post/top-10-benefits-of-reading-for-all-ages/.
19. Aditi Shrikant and Renée Onque, "These Women Use True Crime Podcasts to Lull Themselves to Sleep—Experts Explain Why it Works," CNBC, October 13, 2023, https://www.cnbc.com/2023/10/13/why-are-women-falling-asleep-to-true-crime-podcasts-and-tv-shows.html.
20. "Trespassing is the leading cause of rail-related deaths in the United States. Nationally, more than 500 trespass fatalities occur each year." railroads.dot.gov, April 4, 2025.
21. "Guarding Against the 'Silent Killer,'" CBC News, last modified December 31, 2009, https://www.cbc.ca/news/science/guarding-against-the-silent-killer-1.803889.

Chapter Eight: *Spellbound*

1. Francis Beeding, *The House of Dr. Edwardes* (Boston: Little Brown and Company, 1927), 5.
2. Shae Sennett, "Alfred Hitchcock Gave Ingrid Bergman the Best Acting Advice of Her Career," SlashFilm (website), April 29, 2022, https://www.slashfilm.com/848014/alfred-hitchcock-gave-ingrid-bergman-the-best-acting-advice-of-her-career/.
3. "What is Method Acting?," The Lee Strasberg Theatre & Film Institute, n.d., https://strasberg.edu/about/what-is-method-acting/.
4. Kate Sanders, "Understanding Facial Attractiveness: Exploring the Science Behind Beauty," Reviva Labs (website), July 30, 2023, https://www.revivalabs.com/understanding-facial-attractiveness-exploring-the-science-behind-beauty/.
5. "5 Things You Might Not Know About Alfred Hitchcock's 'Spellbound,'" The Playlist (website), October 31, 2012, https://theplaylist.net/5-things-you-may-not-know-about-alfred-hitchcocks-spellbound-20121031/.
6. Stephanie Leong, Wendi Waits, and Carroll Diebold, "Dissociative Amnesia and DSM-IV-TR Cluster C Personality Traits," *Psychiatry (Edgmont)* 3, no. 1 (January 2006): 51–55, https://pmc.ncbi.nlm.nih.gov/articles/PMC2990548/.
7. Keri Wiginton, "Signs You're Mentally Exhausted," WebMD, reviewed March 21, 2024, https://www.webmd.com/mental-health/ss/slideshow-signs-youre-mentally-exhausted.
8. Marissa Laliberte, "30 Unusual Phobias You Never Knew Existed," The Healthy (website), last modified November 4, 2020, https://www.thehealthy.com/mental-health/strange-phobias/.
9. "Phobias," Johns Hopkins Medicine, n.d., https://www.hopkinsmedicine.org/health/conditions-and-diseases/phobias.

10. "Guilt Complex: Causes, Symptoms, and 8 Ways to Recognize It," Makin Wellness (website), last modified August 27, 2024, https://www.makin-wellness.com/guilt-complex/.

11. Jay Vera Summer, "What Are Precognitive (Premonition) Dreams?," Sleep Foundation (website), last modified January 31, 2023, https://www.sleepfoundation.org/dreams/precognitive-dreams.

12. Summer, "What Are Precognitive Dreams?"

Chapter Nine: *Strangers on a Train*

1. Sven Mikulec, "*Strangers on a Train:* A Technically Perfect Psychological Carousel as One of Hitchcock's Best," Cinephilia & Beyond (website), n.d., https://cinephiliabeyond.org/strangers-train-technically-perfect-psycho-logical-carousel-one-hitchcocks-best/.

2. Dwyer Murphy, "Why did Raymond Chandler Hate *Strangers on a Train* so Intensely?," CrimeReads (website), February 18, 2021, https://crimereads.com/why-did-raymond-chandler-hate-strangers-on-a-train-so-intensely/.

3. Patricia Hitchcock O'Connell, "Patricia Hitchcock O'Connell on Life with Father: Alfred Hitchcock," interview by Joe Leydon, The Moving Picture Show with Joe Leydon (website), March 11, 1993, https://www.movingpictureshow.com/?p=9359.

4. Truffaut, *Hitchcock/Truffaut*, 197.

5. Selina Hurley, "The Surprisingly Old Story of London's First Ever Electric Taxi," *Science Museum* (blog), July 9, 2012, https://blog.sciencemuseum.org.uk/the-surprisingly-old-story-of-londons-first-ever-electric-taxi/.

6. Liam Mays, Shawn Baldwin, and Jeniece Pettitt, "New York City Taxis Battle Uber and Lyft for Riders," CNBC, July 15, 2023, video, 14:39, https://www.cnbc.com/video/2023/07/15/new-york-city-taxis-battle-uber-and-lyft-for-riders.html.

7. "Gait Disorders and Abnormalities," Cleveland Clinic, reviewed February 22, 2023, https://my.clevelandclinic.org/health/diseases/21092-gait-disorders.

8. Becky Little, "When Cigarette Companies Used Doctors to Push Smoking," History.com, last modified April 15, 2025, https://www.history.com/articles/cigarette-ads-doctors-smoking-endorsement.

9. Jonathan W. Roberti, "A Review of Behavioral and Biological Correlates of Sensation Seeking," *Journal of Research in Personality* 38, no. 3 (June 2004): 29, https://doi.org/10.1016/S0092-6566(03)00067-9.

10. Christal Yuen, "How to Fall Asleep in 10, 60, or 120 Seconds," Healthline, last modified February 1, 2024, https://www.healthline.com/health/healthy-sleep/fall-asleep-fast.

11. Marni Feuerman, "What Does 'Mama's Boy' Mean?," Very Well Mind (website), last modified May 13, 2024, https://www.verywellmind.com/ways-to-handle-mamas-boy-husband-4050817.

12. W. J. Camara, J. S. Nathan, and A. E. Puente, "Psychological Test Usage: Implications in Professional Psychology," *Professional Psychology: Research and Practice* 31, no. 2 (2000): 141–54, https://doi.org/10.1037/0735-7028.31.2.141.

13. Josh Victor, "History of Carnival Games," Our Pasttimes (website), April 12, 2017, (website discontinued).

14. Aaron M. White, "What Happened? Alcohol, Memory Blackouts, and the Brain," *Alcohol Research & Health* 27, no. 2 (February 2003): 186–96, https://www.researchgate.net/publication/8406778_What_Happened_Alcohol_memory_blackouts_and_the_brain.

15. James Rotton and I. W. Kelly, "Much Ado About the Full Moon: A Meta-Analysis of Lunar-Lunacy Research," *Psychological Bulletin* 97, no. 2 (March 1985): 286, https://doi.org/10.1037/0033-2909.97.2.286.

16. "Watch as a Big Octopus Squeezes Through a Tiny Hole on a Boat," Texas Hill Country (website), video, 2:07, https://texashillcountry.com/octopus-squeezes-tiny-hole/.

17. Joan Morris, "Morris: Big Creatures Fit Through Tiny Openings," *Mercury News*, last modified August 12, 2016, https://www.mercurynews.com/2013/11/27/morris-big-creatures-fit-through-tiny-openings/.

18. "Accidents in Theme Parks: Unveils the Rollercoaster of Risks." LezDoTechMed (website), December 12, 2024, video, 1:51, https://www.lezdotechmed.com/blog/accidents-in-theme-parks/.

Chapter Ten: *Dial M For Murder*

1. McGilligan, *Alfred Hitchcock*, 475.

2. Elaine Lennon, "Alfred Hitchcock & Grace Kelly," *Offscreen*, May 2017, https://offscreen.com/view/hitchcock-kelly.

3. "Grace Kelly in Dial M for Murder (1954)," *Cinema Cities* (blog), November 12, 2016, https://cinema-cities.com/2016/11/12/grace-kelly-in-dial-m-for-murder-1954/.

4. Mehruss Jon Ahi and Armen Karaoghlanian, "Dial M for Murder (1954)," *Interiors*, 2013, https://www.intjournal.com/1013/dial-m-for-murder.

5. (2006) "10+1 Things You May Not Know About Salt." *Epikouria*.

6. Cunningham, Scott. (1989) *Wicca: A Guide For the Solitary Practitioner*. Llewellyn Worldwide.

7. Nsikan Akpan, "There is No 'Gay Gene.' There is No 'Straight Gene.' Sexuality is Just Complex, Study Confirms," PBS News, August 29, 2019, https://www.pbs.org/newshour/science/there-is-no-gay-gene-there-is-no-straight-gene-sexuality-is-just-complex-study-confirms.

8. "Understanding Sexual Orientation and Homosexuality," American Psychological Association, 2008, https://www.apa.org/topics/lgbtq/orientation.

9. *New World Encyclopedia*, s.v. "Public Broadcasting Service," last modified May 28, 2021, https://www.newworldencyclopedia.org/entry/Public_Broadcasting_Service.

Chapter Eleven: *Rear Window*

1. Lou Lumenick, "Inside the Real Greenwich Village Apartment That Inspired 'Rear Window,'" *New York Post*, August 6, 2014, videos, 2:57, 3:17, https://nypost.com/2014/08/06/inside-the-real-greenwich-village-apartment-that-inspired-rear-window/.
2. Eugene P. Wightman, "Journal of Photography of the George Eastman House, Vol IV, No. 7, October 1955.https://web.archive.org/web/20140809215933/ http://image.eastmanhouse.org/files/GEH_1955_04_07.pdf.
3. Cornell Woolrich, "It Had to Be Murder," *Dime Detective Magazine*, February 1942, 15.
4. Oliver Tearle, "A Summary and Analysis of 'Through a Window' by H.G Wells," Interesting Literature (website), n.d., https://interestingliterature.com/2023/10/hg-wells-through-a-window-summary-analysis/.
5. Truffaut, Francois. (1983) *Hitchcock/Truffaut*. Simon and Schuster.
6. Jason Hellerman, "What Is Pudovkin Montage Theory?," No Film School (website), September 20, 2023, https://nofilmschool.com/pudovkin-montage-theory.
7. "Protecting New Yorkers from Extreme Heat," NYC.gov, September 9, 2020, https://a816-dohbesp.nyc.gov/IndicatorPublic/data-stories/heat/.
8. "When Was the Camera Invented? Everything You Need to Know," Nashville Film Institute, n.d., https://www.nfi.edu/when-was-the-camera-invented/.
9. "When Was Camera Invented?," Nashville Film Institute.
10. Jonathan M. Metzal, "Voyeur Nation? Changing Definitions of Voyeurism, 1950–2004," *Harvard Review of Psychology* 12, no. 2 (2004): 127, https://doi.org/10.1080/10673220490447245.
11. Tianna Soto, "Do Opposites Really Attract? 9 Pros & Cons to Consider," Mind Body Green (website), March 26, 2023, https://www.mindbodygreen.com/articles/do-opposites-really-attract.
12. Maya Boyer, "Women's Intuition: How and Why the Stigmatization of Intuitive Thought Puts Women at Risk" (paper, English 1010, School of Liberal Arts, Tulane University, December 7, 2021), 1, https://liberalarts.tulane.edu/sites/default/files/sites/default/files/6531/Maya%20Boyer%20-%20Womens%20Intuition.pdf.
13. "How Eyes See at Night," *Contact Lenses & Vision* (blog), Cooper Vision, February 14, 2023, https://coopervision.com/blog/how-eyes-see-night.

Chapter Twelve: *Alfred Hitchcock Presents*

1. McGilligan, *Alfred Hitchcock*, 515.
2. McGilligan, *Alfred Hitchcock*, 523.
3. "*Alfred Hitchcock Presents*: Trivia," IMDb, n.d., comment, https://www.imdb.com/title/tt0047708/trivia/?item=tr0772429&ref_=ext_shr_lnk.
4. "*Alfred Hitchcock Presents* Theme: Charles Gounod," Classic FM (website), https://www.classicfm.com/discover-music/periods-genres/film-tv/alfred-hitchcock-presents-theme/ (page removed).

5. Debopriyaa Dutta, "How *Alfred Hitchcock Presents* brought the Director's Filmmaking Sensibilities to the Small Screen," SlashFilm (website), December 17, 2023, https://www.slashfilm.com/1464191/how-alfred-hitchcock-presents-brought-filmmaking-to-small-screen/.

6. "Death Certification," Patient (website), last modified December 15, 2021, https://patient.info/doctor/death-recognition-and-certification.

7. Rijen Shrestha, Tanuj Kanchan, and Kewal Krishan, "Methods of Estimation of Time Since Death," in *StatPearls*, ed. Mohamed Adelsattar et al. (Treasure Island, FL: StatPearls Publishing, January 2025), https://www.ncbi.nlm. nih.gov/books/NBK549867/.

8. Andrew Campbell, "How Mushrooms Are Grown," Canadian Food Focus (website), n.d., video, 1:34, https://canadianfoodfocus.org/on-the-farm/how-mushrooms-are-grown/.

9. Shelby Simon, "Midlife Crisis: Signs, Causes and Treatments," *Forbes*, last modified September 29, 2023, https://www.forbes.com/health/mind/midlife-crisis/.

10. *Oregon Encyclopedia*, s.v. "Humongous Fungus," by Jim Scheppke, last modified October 8, 2024, https://www.oregonencyclopedia.org/articles/humongous-fungus-armillaria-ostoyae/.

11. "This Edible Mushroom Grows in Human Bodies," Learn Your Land (website), https://learnyourland.com/this-edible-mushroom-grows-in-human-bodies/.

12. Marsha M. Linehan, *DBT Skills Training Handouts and Worksheets*, 2nd ed. (New York: Guildford Press, 2015), 20A.

13. Lawrence Kutner, "Midnight Monsters and Imaginary Companions," Psych Central, May 17, 2016, https://psychcentral.com/lib/midnight-monsters-and-imaginary-companions#1.

14. Bambi Turner, "10 Superstitions About Birds," How Stuff Works (website), last modified August 30, 2023, https://animals.howstuffworks.com/birds/10-superstitions-about-birds.htm.

15. Michelle Megna, "Pet Ownership Statistics 2025," *Forbes*, last modified January 2, 2025, https://www.forbes.com/advisor/pet-insurance/pet-ownership-statistics/.

16. Jeff Kunerth, "Pets Are Not People," UCF Forum, University of Central Florida, August 19, 2020, https://stars.library.ucf.edu/cgi/viewcontent.cgi?article=1415&context=ucf-forum.

Chapter Thirteen: *Vertigo*

1. Ty Burr, "Vertigo Is Still the Best Movie Ever. Or the Worst Movie Ever. Discuss," *Washington Post*, May 9, 2023, https://www.washingtonpost.com/movies/2023/05/09/vertigo-at-65-hitchcock-weirdest-movie/.

2. Simon Hattenstone, "'I Had to Leave Hollywood to Save Myself:' Kim Novak on Art, Bipolar, Hitchcock, and Happiness," *The Guardian*, February 15,

2021, https://www.theguardian.com/film/2021/feb/15/i-had-to-leave-hollywood-to-save-myself-kim-novak-on-art-bipolar-hitchcock-and-happiness.

3. Jake Coyle, "60 Years Later, Kim Novak Reflects on 'Vertigo.'" Associated Press, *Chicago Tribune*, March 16, 2018, https://www.chicagotribune.com/2018/03/16/60-years-later-kim-novak-reflects-on-vertigo/.

4. Mike D'Angelo, "France's Answer to Hitchcock Earned the Comparisons with an All-Time Great Thriller," AV Club, October 20, 2020, https://www.avclub.com/france-s-answer-to-hitchcock-earned-the-comparisons-wit-1845396770.

5. Joseph V. Campellone, "Vertigo," Penn Medicine, n.d., https://www.pennmedicine.org/for-patients-and-visitors/patient-information/conditions-treated-a-to-z/vertigo.

6. "Acrophobia (Fear of Heights)," Cleveland Clinic, reviewed October 22, 2021, https://my.clevelandclinic.org/health/diseases/21956-acrophobia-fear-of-heights.

7. Ian Sample, "How Big a Fall Can a Person Survive?," *The Guardian*, May 20, 2004, https://www.theguardian.com/science/2004/may/20/thisweekssciencequestions2.

8. "What is Exposure Therapy?," American Psychological Association, 2017, https://www.apa.org/ptsd-guideline/patients-and-families/exposure-therapy.

9. Longkumer, Imyarila et al. "Social Stigmitization and Late Treatment of Dissociative Disorder: A Case Rport on Trance and Possession Disorder," *Cureus*, December 8, 2023.

10. Matt Miksa, "12 Mind-Blowing Facts from Scientists who Study Reincarnation," MattMiska.com, October 17, 2021, https://mattmiksa.com/blog/12-mind-blowing-facts-from-scientists-who-study-reincarnation.

11. Miksa, "12 Mind-Blowing Facts."

12. Catrina Prager, "Do Blondes Have More Fun? Here's What Science Has to Say," Medium (website), November 4, 2022, https://catrinaprager.medium.com/do-blondes-have-more-fun-heres-what-science-has-to-say-b2325bd879b2.

13. (2024) "Blond Hair Percentage by Country." *World Population Review*.

14. Mount Sinai, "Largest Genetic Study of Suicide Attempts Confirms Genetic Underpinnings That Are Not Driven by Underlying Psychiatric Disorders," news release, November 29, 2021, https://www.mountsinai.org/about/newsroom/2021/largest-genetic-study-of-suicide-attempts-confirms-ge-netic-underpinnings-that-are-not-driven-by-underlying-psychiatric-dis-orders.

15. Laura Moss, "Why Can We Sense When People Are Looking at Us?," Treehugger (website), last modified August 24, 2024, https://www.tree-hugger.com/why-can-we-sense-when-people-are-looking-us-4864192.

16. "Giant Sequoia Facts," Save the Redwoods League (website), n.d., https://www.savetheredwoods.org/redwoods/giant-sequoias/.

Chapter Fourteen: North By Northwest

1. Sven Mikulec, "'North by Northwest': Quite Possibly the Most Entertaining Hitchcock Ever," Cinephilia & Beyond, n.d., https://cinephiliabeyond.org/north-by-northwest-quite-possibly-the-most-entertaining-hitchcock-ever/.

2. Lang, Brett. (July 4, 2021) "As Eva Marie Saint Turns 97, Celebrating her Seductive Turn in North by Northwest." Variety.

3. "The Intoxication Defense in Criminal Law Cases," Justia (website), n.d., https://www.justia.com/criminal/defenses/intoxication/.

4. William Mattar Law Offices, "The Science of Drunk Driving," Our Blog, April 26, 2023, https://williammattar.com/blog/drunk-driving/the-science-of-drunk-driving/?utm_content=organic_direct.

5. People of the State of Illinois v. William "Duff" Armstrong, Cir. Sangamon Cty. (1858), Illinois State Bar Association, https://www.isba.org/sites/default/files/teachers/mocktrial/2009%20Lincoln%20Almanac%20Trial.pdf.

6. Eric Rasmussen, "Mistaken Identity Case Resolved After 5 Investigates Exposes Error in Court Records," KSTP.com, September 19, 2023, https://kstp.com/kstp-news/top-news/mistaken-identity-case-resolved-after-5-investigates-exposes-error-in-court-records/.

7. "History of Women's Shaving." Shaving & Grooming Blog, The Razor Company, April 29, 2023, https://www.therazorcompany.com/blogs/history-of-wet-shaving/history-of-womens-shaving?srsltid=AfmBOoof-RBQXKrARaGid-9ZzS2IRTlV5ciDDv1sUsR859VdUZemMljoF.

8. SkyQuest, Global Shaving Market Size, Share, and Growth Analysis (Westford, MA: SkyQuest, January 2025), https://www.skyquestt.com/report/shaving-market.

9. David Miller, "Hitchcock's Most Ridiculed Scene Actually Made Perfect Sense," Screen Rant (website), May 6, 2022, https://screenrant.com/north-by-northwest-hitchcock-crop-duster-scene-logical/.

10. "The Psychology of Bidding at Auction." Dakil Auctioneers, n.d., https://www.dakil.com/auction-101/psychology-bidding-auction.

11. Esther Inglis-Arkell, "Why a Gun Loaded With Blanks Can Still Kill You," Gizmodo (website), October 27, 2021, https://gizmodo.com/why-a-gun-loaded-with-blanks-can-still-kill-you-5972313.

12. Stephen Mays, "100 National Park Service Facts," U.S. News & World Report, August 19, 2016, https://www.usnews.com/news/articles/2016-08-19/100-national-park-service-facts.

13. Jennifer Nalewicki, "Five Colossal Stone Portraits Around the World," Smithsonian, November 30, 2016, https://www.smithsonianmag.com/travel/six-massive-carved-monuments-you-need-see-believe-180960795/.

Chapter Fifteen: *Psycho*

1. Meg Hafdahl and Kelly Florence, *The Science of Monsters: The Truth about Zombies, Witches, Werewolves, Vampires, and Other Legendary Creatures* (New York: Skyhorse Publishing, 2019), 21.
2. "The Backstory to Robert Bloch's *Psycho*," Galaxy Press, n.d., https://galaxypress.com/backstory-to-psycho/.
3. Truffaut, *Hitchcock/Truffaut*, 269.
4. Mike Floorwalker, "Why Janet Leigh Was Never the Same After Psycho," Looper (website), https://www.looper.com/182476/why-janet-leigh-was-never-the-same-after-psycho/.
5. Rachel Heston-Davis, "The History of Anxiety," Psych Central (website), last modified April 24, 2023, https://psychcentral.com/anxiety/the-origins-of-anxiety.
6. Heston-Davis, "History of Anxiety."
7. Rick Paulas, "The Psychological Impact of Driving Among Police Cars," Pacific Standard (website), September 29, 2015, https://psmag.com/news/bad-boys-bad-boys-what-you-gonna-do/.
8. "New Poll Reveals the Psychology Behind Post-Pandemic Car Buyers." *True Car Blog*, January 9, 2023, https://www.truecar.com/blog/new-poll-reveals-the-psychology-behind-post-pandemic-car-buyers/.
9. Lyndsey Matthews, "Please Don't Put Your Suitcase on the Bed, Ever," *AFAR*, March 14, 2024, https://www.afar.com/magazine/why-you-should-never-put-your-suitcase-on-the-bed.
10. Sara Thomas, "How Common are Bed Bugs in Hotels?," Hotel Chantelle (website), last modified August 1, 2023, https://hotelchantelle.com/how-common-are-bed-bugs-in-hotels/.
11. "Standing Up and Stepping In," Guardian Life Insurance Company, n.d., https://www.guardianlife.com/reports/caregiving-in-america.
12. "How Can I Take Time Off Work to Care for a Family Member?," Family Caregiver Alliance, n.d., https://www.caregiver.org/faq/how-can-i-take-time-off-work-to-care-for-a-family-member/.
13. "What are Five Tips for Escaping a Sinking Car?," Foster Wallace, n.d., https://www.fosterwallace.com/blog/five-tips-for-escaping-a-sinking-car/.
14. Lara James, "Oral Fixation: Habits, Risks, and Resolutions." Dentistry IQ (website), October 26, 2022, https://www.dentistryiq.com/dentistry/article/14284820/oral-fixation-habits-risks-and-resolutions.
15. "Causes of Stuttering," National Stuttering Association, n.d., https://www.westutter.org/post/causes-of-stuttering.

Chapter Sixteen: *The Birds*

1. McGilligan, *Alfred Hitchcock*, 613.
2. Hafdahl and Florence, *Science of Monsters*, 121.

3. "Birds," Centers for Disease Control, n.d., https://www.cdc.gov/healthy-pets/about/birds.html.

4. McGilligan, *Alfred Hitchcock*, 615.

5. Tom Warren, "Can Birds Predict the Weather?," *The Old Farmer's Almanac*, last modified January 14, 2025, https://www.almanac.com/how-birds-predict-weather.

6. Tim Falk, "Bird Care 101: A Primer for New Pet Parents," Wag! (website), last modified March 30, 2023, https://wagwalking.com/wellness/bird-care-101-a-primer-for-new-pet-parents.

7. Paul Germain, Monique Piau, and Denis Caillerlie, eds., *Theoretical and Applied Mechanics* (Amsterdam: Elsevier Science Publishers, 1989).

8. Leon James, "Psychology Behind the Wheel: Why Do We Speed?," interview by Neal Conan, NPR, June 7, 2007, https://www.npr.org/transcripts/10812153.

9. James, "Psychology Behind the Wheel."

10. Jane Smith, "Bird Attacks: How Many Humans Do Birds Kill Each Year?," Fauna Advice (website), September 7, 2023, (website discontinued).

11. Niki Yuen Fen Bruysters and Pamela D. Pilkington, "Overprotective Parenting Experiences and Early Maladaptive Schemas in Adolescence and Adulthood: A Systematic Review and Meta-Analysis," *Clinical Psychology & Psychotherapy* 30, no. 1 (January-February 2023): 12, https://doi.rog/10.1002/cpp.2776.

12. Julie Rach Mancini, *Why Does My Bird Do That? A Guide to Parrot Behavior*, 2nd ed. (Hoboken, NJ: Wiley Publishing, 2006), 183.

13. Greg Llopis, "How Great Leaders Maintain Their Composure Under Intense Pressure," Leadership Preview (website), n.d., https://www.leadershippreview.net/how-great-leaders-maintain-their-composure-under-intense-pressure/.

14. Douglas Main, "How Many Birds Are There in the World?," *National Geographic*, May 17, 2021, https://www.nationalgeographic.com/animals/article/how-many-birds-are-there-in-the-world-science-estimates.

15. Nathan J. Emery, "Cognitive Ornithology: The Evolution of Avian Intelligence," *Philosophical Transactions of the Royal Society B* 361, no. 1465 (January 29, 2006): 23, https://doi.org/10.1098/rstb.2005.1736.

Chapter Seventeen: *Marnie*

1. McGilligan, *Alfred Hitchcock*, 605.

2. Eugene Archer, "Hitchcock's *Marnie* with Tippi Hedren and Sean Connery," *New York Times*, July 23, 1964, https://archive.nytimes.com/www.nytimes.com/library/film/072364hitch-marnie-review.html.

3. Richard Brody, "*Marnie* Is the Cure for Hitchcock Mania," *New Yorker*, August 17, 2016, https://www.newyorker.com/culture/richard-brody/marnie-is-the-cure-for-hitchcock-mania.

4. "Chromophobia (Fear of Colors)," Cleveland Clinic, reviewed March 22, 2022, https://my.clevelandclinic.org/health/diseases/22580-chromophobia-fear-of-colors.

5. Lindsay Dodgson, "7 Science-Backed Reasons Why You May Be Better Off Being Single," *Business Insider*, February 23, 2023, https://www.businessinsider.com/reasons-youre-better-off-being-single-according-to-science-2019-2.

6. Bruce Y. Lee, "57% of Single Adults in U.S. Not Looking to Date, Survey Says," *Forbes*, February 13, 2023, https://www.forbes.com/sites/brucelee/2023/02/13/for-valentines-day-5-survey-findings-about-singles-from-pew-research/.

7. "8 Ways to Improve Your Powers of Observation," Mindtools (website), n.d., https://www.mindtools.com/acjxune/8-ways-to-improve-your-powers-of-observation.

8. "Memorizing Dates and Numbers," Archimedes Lab Project, accessed January 11, 2024, https://www.archimedes-lab.org/memorizing_numbers.html.

9. Chelsea Whyte, "The 7 Non-Human Mammals Where Females Rule the Roost," *New Scientist*, September 26, 2018, https://www.newscientist.com/article/2180434-the-7-non-human-mammals-where-females-rule-the-roost/.

10. "Kleptomania," Mayo Clinic, September 30, 2022, https://www.mayoclinic.org/diseases-conditions/kleptomania/symptoms-causes/syc-20364732.

11. LGBTQ Center, "Asexuality, Attraction, and Romantic Orientation" University of North Carolina at Chapel Hill, n.d., https://lgbtq.unc.edu/resources/exploring-identities/asexuality-attraction-and-romantic-orientation/.

12. "Dissociation and Trauma in Young People," Orygen, the National Centre of Excellence in Your Mental Health, 2018, https://orygen.org.au/Training/Resources/Trauma/Fact-sheets/Dissociation-trauma/Orygen_Dissociation_and_trauma_in_young_people_fac?ext=..

13. Stephanie Pappas, "Should You Wake Someone from the Throes of a Nightmare?," *Scientific American*, October 5, 2023, https://www.scientificamerican.com/article/should-you-wake-someone-from-the-throes-of-a-nightmare/.

14. Sosnoski, Karen. (October 7, 2021) "All About Free Association Therapy." *Psych Central.com*.

15. John Cherwa, "Why Do Race Horses Keep Dying? Inside the Sport's Push to Solve a Formidable Problem," *Los Angeles Times*, June 9, 2023, https://www.latimes.com/sports/story/2023-06-09/why-race-horses-die-at-the-track.

16. Cherwa, "Why Do Race Horses Keep Dying?"

17. Suzanne Lego, "Repressed Memory and False Memory," *Archives of Psychiatric Nursing* 10, no. 2 (April 1996): 110–15, https://doi.org/10.1016/s0883-9417(96)80073-2.

Chapter Eighteen: *Torn Curtain*

1. McGilligan, *Alfred Hitchcock*, 664.

2. Swapnil Dhruv Bose, "Revisiting the Time Paul Newman Clashed with Alfred Hitchcock," Far Out Magazine, December 29, 2022, https://farout-magazine.co.uk/paul-newman-clashed-with-alfred-hitchcock/.

3. Steve Vertlieb, "Herrmann and Hitchcock: The Torn Curtain," Bernard Herrmann Society (website), 2002, http://bernardherrmann.org/articles/misc-torncurtain/index.html.

4. Kathryn Watson, "How to Increase Your Body Temperature," Healthline, January 5, 2021, https://www.healthline.com/health/how-to-increase-body-temperature.

5. Marc Silver, "Pi Day Turns 25: Why We Celebrate an Irrational Number," *National Geographic*, March 14, 2013, https://www.nationalgeographic.com/science/article/031314-pi-day-exploratorium-mathematics-pie-science.

6. "Hazards of Natural Gas," Pinedale Natural Gas, n.d., https://pinedalegas.com/pipline-safety/hazards-of-natural-gas/.

7. Joseph Nye, "A Generation Later, Revisiting the Debate on Nuclear Weapons Ethics," interview by James F. Smith, Harvard Kennedy School, May 26, 2023, https://www.hks.harvard.edu/faculty-research/policy-topics/international-relations-security/generation-later-revisiting-debate.

8. Nye, "A Generation Later."

9. Anne Marie Helmenstine, "Why Mathematics Is a Language," ThoughtCo (website), last modified June 27, 2019, https://www.thoughtco.com/why-mathematics-is-a-language-4158142.

10. Carlton F. W. Larson, "Shouting 'Fire' in a Theater: The Life and Times of Constitutional Law's Most Enduring Analogy," *William & Mary Bill of Rights Journal* 24, no. 1 (October 2015): 187, https://scholarship.law.wm.edu/cgi/viewcontent.cgi?article=1748&context=wmborj.

Chapter Nineteen: Family Plot

1. Eric B. Jones, *Family Plot*—Classic Reaction," Medium (website), November 22, 2021, https://medium.com/film-cut/family-plot-classic-reaction-6bae33fa2783.

2. Tom Shales, "Alfred Hitchcock, Master of Screen Thrillers, Dies," *Washington Post*, April 29, 1980, https://www.washingtonpost.com/archive/local/1980/04/30/alfred-hitchcock-master-of-screen-thrillers-dies/0f18e689-1868-4a3d-8b97-de2271807b51/.

3. Neil Dagnall and Ken Drinkwater, "The Science of Why So Many People Believe in Psychic Powers," The Conversation (website), February 4, 2019, https://theconversation.com/the-science-of-why-so-many-people-believe-in-psychic-powers-102088.

4. Dagnall and Drinkwater, "Science of Why So Many People Believe."

5. American Academy of Pediatrics, "Hearing and Making Sounds: Your Baby's Milestones," Healthy Children (website), last modified February 22, 2021, https://www.healthychildren.org/English/ages-stages/baby/Pages/Hearing-and-Making-Sounds.aspx.

6. Sondra Francis, "The Four Cs of Gemstone Valuation," Ganoksin (website), last modified October 30, 2023, https://www.ganoksin.com/article/4-cs-gemstone-valuation/.

7. Drug Enforcement Administration, "Ketamine," Drug Fact Sheet, April 2020, https://www.dea.gov/sites/default/files/2020-06/Ketamine-2020.pdf.

8. Edgar Allan Poe, "The Purloined Letter," in *The Works of Edgar Allan Poe* (New York: Harper and Bros., 1910), 59.

9. (2024) "*Family Plot*: Trivia," IMDb, n.d., comment, https://www.imdb.com/title/tt0074512/trivia/?ref_=tt_dyk_trv.

10. Sandee LaMotte, "Cremation Has Replaced Traditional Burials in Popularity in America and People Are Getting Creative with Those Ashes," CNN, last modified January 23, 2020, https://www.cnn.com/2020/01/22/health/cremation-trends-wellness/index.html.

11. Jill Darby, "The History of Grave Markers in America," Trigard Memorials (website), February 22, 2023, https://www.trigardmemorials.com/blog/the-history-of-grave-markers-in-america/.

12. Harvard Health Publishing, "Tips to Retrieve Old Memories," Harvard Medical School, April 1, 2021, https://www.health.harvard.edu/mind-and-mood/tips-to-retrieve-old-memories.

13. "What to Do If Your Brakes Go Out," Driver Ed Safety (website), February 12, 2021, https://driveredsafety.com/what-if-your-brakes-go-out/.

14. Ned Herrmann, "What Is the Function of the Various Brainwaves?," *Scientific American*, December 22, 1997, https://www.scientificamerican.com/article/what-is-the-function-of-t-1997-12-22/.

15. Krishna Maxwell, "Playing Possum: 9 Animals That Play Dead to Survive." AZ Animals (website), June 3, 2023, [URL] (webpage removed).

Conclusion

1. Madeline Di Nonno, in discussion with the authors, August 29, 2023.

Index